Algebra II Topics by Design

Russell F. Jacobs

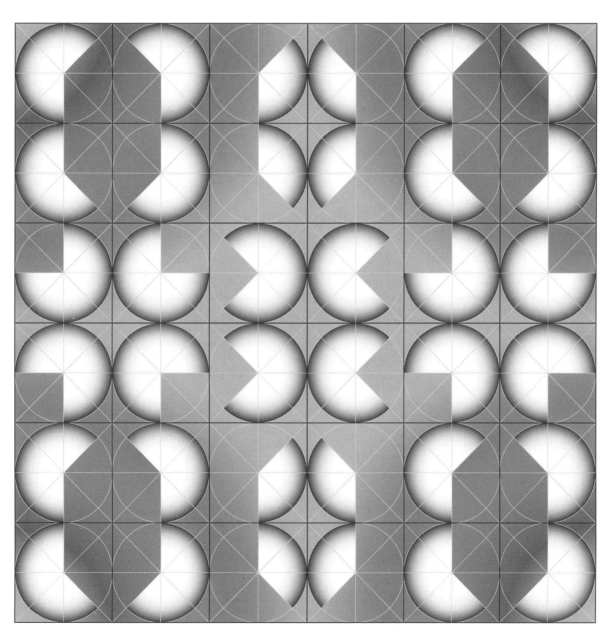

Jacobs Publishing

An imprint of Tessellations

Phoenix, Arizona

AUTHOR   Russell F. Jacobs

Editorial Assistant   Victor Bobbett, Design and Composition

Consultants   John Rucker, Alan Jacobs, Sally Jacobs, Ph. D.

To the Teacher

The following directions may be helpful in getting students started with **Algebra II by Design**.

"Work each exercise and search for the answer on the grid. Each answer will appear one or more times. Shade each square containing the answer just like the small square or circle next to the exercise is shaded. If the grid is shaded correctly, a pleasing design emerges."

Both the student and the teacher usually can tell at once whether or not the work is correct. Most of the designs are symmetrical. Any errors in working the exercises or in shading are noticeable. For convenience, an **Answer Key** for all the activities is included in the back of the book.

Some students may want to create designs in color. It is recommended that they first do the shading with a lead pencil to make sure the design is correct. Then they can redo the design in color on a separate grid using a coloring pen or pencil.

The exercises in **Algebra II by Design** are for review and practice of certain skills and concepts usually covered in a second year algebra course. In most cases there is no explanation of how the exercises are to be worked. However, in some exercises an example is shown. The activities are to be used to supplement the regular mathematics curriculum.

Special Note

The activities, and even exercises within an activity, are of varying degrees of difficulty. Several activities can be viewed as review of Algebra One concepts. Some activities may be too challenging for some students. You should examine each activity before assigning it to determine whether or not it is appropriate for your students.

ISBN 0-918272-38-6

Table of Contents

Note: The letters and numbers in parentheses following the activity name reference specific Common Core State Standards for Mathematics addressed by the activity

Solution Key (See pages following Activity 42.)

Activity 1

Name: _____

	A	B	C	D	E	F	
	$-\frac{2}{5}$	$\frac{1}{4}$	1	$-\frac{3}{7}$	$-\frac{2}{5}$	$\frac{1}{4}$	1
	$-\frac{5}{3}$	0	$-\frac{4}{3}$	$-\frac{1}{2}$	$-\frac{5}{3}$	0	2
	-2	4	$\frac{3}{2}$	$-\frac{5}{2}$	-2	4	3
	$\frac{1}{2}$	3	$-\frac{5}{2}$	$\frac{3}{2}$	$\frac{1}{2}$	3	4
	$-\frac{2}{5}$	$\frac{1}{4}$	1	$-\frac{3}{7}$	$-\frac{2}{5}$	$\frac{1}{4}$	5
	$-\frac{5}{3}$	0	$-\frac{4}{3}$	$-\frac{1}{2}$	$-\frac{5}{3}$	0	6

Find the slope of each graph.

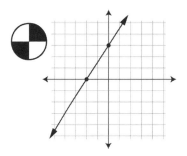

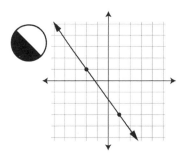

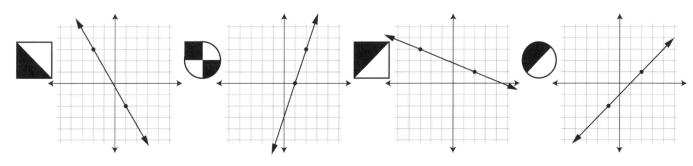

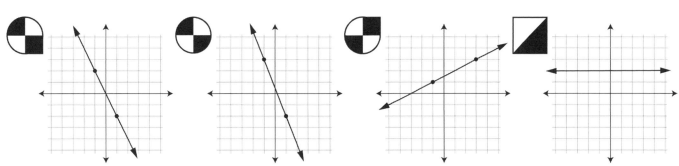

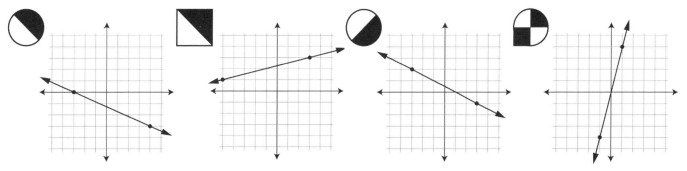

Activity 2   Name: _____

	A	B	C	D	E	F	
1	2	$-\frac{1}{2}$	$\frac{1}{3}$	5	2	$-\frac{1}{2}$	1
2	$\frac{1}{2}$	$-\frac{3}{2}$	$y=-2x+6$	-3	$\frac{1}{2}$	$-\frac{3}{2}$	2
3	$y=-\frac{2}{3}x-1$	$y=5x-2$	-2	3	$y=-\frac{2}{3}x-1$	$y=5x-2$	3
4	$y=3x+\frac{1}{4}$	$\frac{3}{2}$	$y=\frac{1}{2}x-3$	$y=x-\frac{3}{2}$	$y=3x+\frac{1}{4}$	-4	4
5	2	$-\frac{1}{2}$	$\frac{1}{3}$	5	2	$-\frac{1}{2}$	5
6	$\frac{1}{2}$	$-\frac{3}{2}$	$y=-2x+6$	4	$\frac{1}{2}$	$-\frac{3}{2}$	6

Find the slope of the graph of each function represented by the given equation.

Example: $y = -\frac{1}{3}x + 2$
Slope $= -\frac{1}{3}$

 $y = -2x + 5$

 $y = \frac{3}{2}x - 2$

 $y = 4 - 3x$

 $x + 2y = 2$ $-3x + y = 4$ $5 = 2y - x$

Find the Y intercept of each function. (Y intercept is the value of y when x equals zero.)

 $y = 3x + 4$ $y = 5 - x$ $x - 2y = 3$

Find the X intercept of each function. (X intercept is the value of x when y equals zero.)

 $y = -2x + 4$ $3x - 3y = 1$ $2x - y + 8 = 0$

Write an equation with the given slope and Y intercept.

 slope $= -2$
Y intercept $= 6$

slope $= \frac{1}{2}$
Y intercept $= -3$

slope $= -\frac{2}{3}$
Y intercept $= -1$

slope $= 5$
Y intercept $= -2$

slope $= 1$
Y intercept $= -\frac{3}{2}$

slope $= 3$
Y intercept $= \frac{1}{4}$

Activity 3   Name: _____

	A	B	C	D	E	F	
	$\frac{1}{2}$	-8	-2	1	$\frac{1}{2}$	-8	1
	$\frac{2}{3}$	$-\frac{3}{8}$	$\frac{3}{4}$	$-\frac{5}{6}$	$\frac{2}{3}$	$-\frac{3}{4}$	2
	$\frac{3}{2}$	$\frac{1}{3}$	1	-2	$\frac{3}{2}$	$\frac{1}{3}$	3
	-3	$-\frac{1}{5}$	$-\frac{5}{6}$	$\frac{3}{4}$	-3	$-\frac{1}{5}$	4
	$\frac{1}{2}$	-8	-2	1	$\frac{1}{2}$	-8	5
	$\frac{2}{3}$	$-\frac{3}{4}$	$\frac{3}{4}$	$-\frac{5}{6}$	$\frac{2}{3}$	$-\frac{3}{8}$	6

Find the slope of the line containing the two points A and B.

 A = (−2,1)   B = (0,3)

 A = (1,4)   B = (−2,3)

 A = (4,2)   B = (−1,3)

 A = (−2,−3)   B = (2,−1)

 A = (5,−2)   B = (−3,4)

 A = (2,−2)   B = (−1,4)

 A = (5,1)   B = (3,−2)

 A = $(3\frac{1}{2},1\frac{1}{4})$ B = $(\frac{1}{2},-\frac{3}{4})$

 A = $(-4,\frac{1}{2})$ B = $(-3,1\frac{1}{4})$

A = (1,2)   B = (0,5)

A = (−4,1)   B = (4,−2)

A = $(-1,-\frac{1}{2})$ B = $(-1\frac{1}{2},3\frac{1}{2})$

A = (−3,4)   B = (3,−1)

Activity 4

Name: _____

Algebra II Topics by Design / Russell F. Jacobs   www.tessellations.com   © 2007 Jacobs Publishing Company

	A	B	C	D	E	F	
1	$3x-4y=-12$	$y=-5x+1.8$	$x+y+2=0$	$y=\frac{1}{3}x+\frac{3}{4}$	$3x-4y=-12$	$y=-5x+1.8$	1
2	$y=-2x-1$	$y=\frac{4}{3}x-2$	$y=-2x-\frac{5}{4}$	$y=-2x+3$	$y=-2x-1$	$y=\frac{4}{3}x-2$	2
3	$y=-\frac{3}{2}x+\frac{2}{3}$	$y=\frac{2}{3}x$	$y=2x-10$	$y=\frac{1}{3}x-\frac{3}{2}$	$y=-\frac{3}{2}x+\frac{2}{3}$	$y=\frac{2}{3}x$	3
4	$y=\frac{1}{3}x-5$	$3x-3y+4=0$	$y=2x-10$	$y=\frac{1}{3}x-\frac{3}{2}$	$y=\frac{1}{3}x-5$	$3x-3y+4=0$	4
5	$3x-4y=-12$	$y=-5x+1.8$	$y=-x$	$y=\frac{1}{3}x+\frac{3}{4}$	$3x-4y=-12$	$y=-5x+1.8$	5
6	$y=-2x-1$	$y=\frac{4}{3}x-2$	$y=-2x-\frac{5}{4}$	$y=-2x+3$	$y=-2x-1$	$y=\frac{4}{3}x-2$	6

Write the equation of a line as described.

 Line is perpendicular to $y = \frac{1}{2}x - 2$ and has Y intercept of 3.

 Line is perpendicular to $y = -3x + 1$ and has Y intercept of -5.

 Line is perpendicular to $y = x + 2$ and has Y intercept of 0.

Line is perpendicular to $y = -\frac{3}{4}x + 3$ and has Y intercept of -2.

Line is perpendicular to $y = 0.2x - 0.5$ and has Y intercept of 1.8.

Line is perpendicular to $y = 4 - 3x$ and has Y intercept of $\frac{3}{4}$.

Line is perpendicular to $x + 2y = 10$ and has Y intercept of -10.

Line is perpendicular to $2x - 3y + 1 = 0$ and has Y intercept of $\frac{2}{3}$.

Line is perpendicular to $2x - 3 = 4y$ and has Y intercept of $-\frac{5}{4}$.

Line is parallel to $y = -2x + 3$ and has Y intercept of -1.

Line is parallel to $y = \frac{3}{4}x - 3$ and has Y intercept of 3.

Line is parallel to $y = -x + \frac{1}{2}$ and has Y intercept of -2.

Line is parallel to $2x - 3y = 1$ and has Y intercept of 0.

Line is parallel to $x = 3y - 2$ and has Y intercept of $-\frac{3}{2}$.

Line is parallel to $x - y + 1 = 0$ and has Y intercept of $\frac{4}{3}$.

Name:_____

	A	B	C	D	E	F	
1	(−1, −1)	(−2, 2)	$(1\frac{3}{4}, -\frac{3}{4})$	(−5, 2)	(−1, −1)	(−5, −5)	1
2	(−1, 4)	$(2\frac{7}{8}, -1\frac{5}{8})$	(−2, −1)	(−5.1, −1.5)	$(-4\frac{1}{4}, 1\frac{1}{2})$	$(2\frac{7}{8}, -1\frac{5}{8})$	2
3	(−5, −5)	(−1, −1)	(−5, −2)	(−3, −2)	(−2, 2)	(−1, −1)	3
4	$(2\frac{7}{8}, -1\frac{5}{8})$	$(-4\frac{1}{4}, 1\frac{1}{2})$	(6, −6)	(6.8, −4.4)	$(2\frac{7}{8}, -1\frac{5}{8})$	(−1, 4)	4
5	(−1, −1)	(−5, −5)	$(1\frac{3}{4}, -\frac{3}{4})$	(−7, −5)	(−1, −1)	(−2, 2)	5
6	(−1, 4)	$(2\frac{7}{8}, -1\frac{5}{8})$	(−2, −1)	(−5.1, −1.5)	$(-4\frac{1}{4}, 1\frac{1}{2})$	$(2\frac{7}{8}, -1\frac{5}{8})$	6

Find the sum of the ordered pairs in each exercise.

(−2, 3) + (−3, −1)

(4, 1) + (−6, −2)

(−10, 3) + (5, −5)

(0, −3) + (−1, 7)

(−6, 3) + (1, −8)

(12, −5) + (−6, −1)

(−9, 3) + (6, −5)

(4, 6) + (−5, −7)

(−6, −3) + (−1, −2)

$(2\frac{1}{2}, -1\frac{1}{4}) + (-\frac{3}{4}, \frac{1}{2})$

$(-3\frac{3}{4}, 2\frac{1}{4}) + (-\frac{1}{2}, -\frac{3}{4})$

$(3\frac{1}{8}, -2\frac{3}{8}) + (-\frac{1}{4}, \frac{3}{4})$

(8.4, −5.6) + (−1.6, 1.2)

(−3.9, 4.3) + (−1.2, −5.8)

$(-1\frac{5}{8}, 2\frac{3}{8}) + (-\frac{3}{8}, -\frac{3}{8})$

Activity 6

Name:_____

	A	B	C	D	E	F	
1	b + a	(a · b)c	a(b + c)	a + (–b)	b + a	(a · b)c	1
2	(a · b)c	$\frac{1}{a}$	–1(a)	b · a	(a · b)c	$\frac{1}{a}$	2
3	(–a) + (–b)	a + (b + c)	0	–(ab)	(–a) + (–b)	a + (b + c)	3
4	a · b + a · c	a($\frac{1}{b}$)	a – b – c	–a	a · b + a · c	a($\frac{1}{b}$)	4
5	b + a	(a · b)c	a(b + c)	a + (–b)	b + a	(a · b)c	5
6	(a · b)c	$\frac{1}{a}$	–1(a)	b · a	(a · b)c	$\frac{1}{a}$	6

Complete each equality by applying the property of real numbers that is given.

 (Commutative Property)

$a \cdot b = $ _____

 (Distributive Property)

$a(b + c) = $ _____

 (Meaning of Subtraction)

$a - b = $ _____

 (Associative Property)

$a(b \cdot c) = $ _____

 (Commutative Property)

$a + b = $ _____

 (Distributive Property)

$a \cdot b + a \cdot c = $ _____

 (Special Property of Multiplication)

$(-a)(b) = $ _____

 (Property of the Opposite of a Sum)

$-(a + b) = $ _____

 (Meaning of Division)

$a \div b = $ _____

 (Additive Inverse)

$a + $ _____ $= 0$

 (Identity Element)

$a + $ _____ $= a$

(Multiplicative Inverse)

$a \cdot $ _____ $= 1$

(Associative Property)

$(a + b) + c = $ _____

 (Multiplication Property of –1)

$-a = $ _____

 (Special Property of Subtraction)

$a - (b + c) = $ _____

Algebra II Topics by Design / Russell F. Jacobs   www.tessellations.com   © 2007 Jacobs Publishing Company

Name:_____

Algebra II Topics by Design / Russell F. Jacobs   www.tessellations.com   © 2007 Jacobs Publishing Company

	A	B	C	D	E	F	
	(−4, 0)	(−4, −1)	(2, 6)	(2, 5)	(−4, 0)	(−4, −1)	1
	(−2, 7)	(−2, −6)	(2, 6)	(2, 5)	(−2, 7)	(−2, −6)	2
	(−8, −4)	(2, −4)	(4, 6)	(6, 6)	(−8, −4)	(2, −4)	3
	(2, 4)	(0, 6)	(1, −7)	(3, −1)	(2, 4)	(0, 6)	4
	(−4, 0)	(−4, −1)	(2, 6)	(2, 5)	(−4, 0)	(−4, −1)	5
	(−2, 7)	(−2, −6)	(2, 6)	(2, 5)	(−2, 7)	(−2, −6)	6

Write an ordered pair that represents the sum of the two vectors.

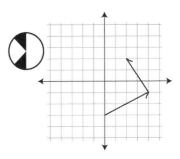

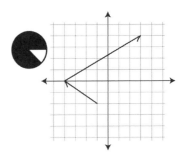

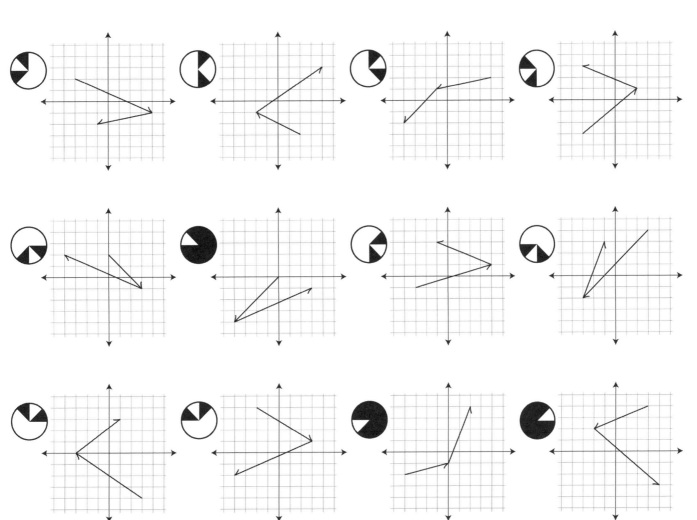

Activity 8

Name: _____

	A	B	C	D	E	F	
1	$-\frac{1}{2}$	$\frac{1}{2}$	$2\frac{1}{2}$	$-\frac{5}{8}$	$-\frac{1}{2}$	$\frac{2}{3}$	1
2	$\frac{2}{3}$	$-\frac{1}{2}$	1	-1	$\frac{1}{2}$	$-\frac{1}{2}$	2
3	$\frac{1}{4}$	-3	$-2\frac{1}{2}$	-2	$\frac{1}{4}$	-3	3
4	$1\frac{2}{3}$	$-4\frac{1}{2}$	$3\frac{1}{2}$	$-\frac{1}{5}$	$1\frac{2}{3}$	$-4\frac{1}{2}$	4
5	$-\frac{1}{2}$	$\frac{2}{3}$	$2\frac{1}{2}$	$-\frac{5}{8}$	$-\frac{1}{2}$	$\frac{1}{2}$	5
6	$\frac{1}{2}$	$-\frac{1}{2}$	1	-1	$\frac{2}{3}$	$-\frac{1}{2}$	6

Solve each equation.

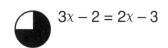 $3x - 2 = 2x - 3$

$3 - 4x = -4 - 2x$

$-5x - 3 = 3 - 3x$

$-2(3 - x) = -(x + 1)$

$2 - (x + 4) = x - 3$

$-2x + 3(-1 + x) = x - (2x + 4)$

$5 - 3(1 - 2x) = x - 3 - 3x$

$x^2 - 2(3 - 2x) = -x(5 - x)$

$x(3 - 2x) = -2x(x - 3) + 6$

$3x - x(4x + 5) = -x(x + 4) - 3x^2 - 5$

$(x^2 - 2) - (x^2 + x) = 10 - (3 - x)$

$\frac{x}{2} - \frac{2x}{5} = \frac{1}{4}$

$-\frac{3x}{4} + \frac{3}{8} = \frac{3}{16}$

$2(1 - x) = 3(-x + 1)$

$\frac{3}{x} - 5 = 5 + \frac{5}{x}$

www.tessellations.com

Name: _____

Solve each inequality.

	A	B	C	D	E	F	
	$x \leq -1$	$x < 6$	$x \geq 4\frac{1}{2}$	$x > 4$	$x > 7$	$x > -6$	1
	$x > -6$	$x > 7$	$x > -2\frac{1}{2}$	$x \leq 1\frac{3}{4}$	$x < 6$	$x \leq -1$	2
	$x < 3$	$x < -4$	$x \leq 2$	$x > -1\frac{1}{4}$	$x < 3$	$x < -4$	3
	$x < 3$	$x < -4$	$x < -6$	$x > \frac{3}{4}$	$x < 3$	$x < -4$	4
	$x > 7$	$x > -6$	$x \geq 4\frac{1}{2}$	$x > 4$	$x \leq -1$	$x < 6$	5
	$x < 6$	$x \leq -1$	$x > -2\frac{1}{2}$	$x \leq 1\frac{3}{4}$	$x > -6$	$x > 7$	6

 $3x + 4 < 13$

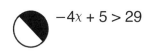 $-4x + 5 > 29$

 $12 < 8 - x$

$-6 + 2x < 6$

$x < 3x - 8$

$-7 + 4x > 3x$

$2x - 4 < 3x + 2$

$3 - x \leq 5 - 2x$

$3x - 2 \geq x + 7$

$-4x - 3 > -8x - 8$

$5x + 6 \leq x + 13$

$3 - (2x - 4) > 2(1 - 2x)$

$3(2 - 3x) - 8 < -6 - (5 - 3x)$

$4(x + 2) + 5 \leq 6 - (2x - 1)$

Activity 10

Name:_____

	A	B	C	D	E	F	
1	2	n^4	$-2n^2$	$\frac{1}{n^2}$	$-6n^6$	9	1
2	$\frac{n^6}{2}$	$n\sqrt{2}$	n^7	n^8	-5	-3	2
3	$-3\sqrt{n}$	n^2	$\frac{1}{n}$	n^6	$-3\sqrt{n}$	n^2	3
4	$\frac{36}{n^6}$	$\frac{1}{3n^2}$	$\frac{1}{n}$	n^6	$\frac{36}{n^6}$	$\frac{1}{3n^2}$	4
5	$-6n^6$	9	$-2n^2$	$4n^2$	2	n^4	5
6	-5	-3	n^7	n^8	$\frac{n^6}{2}$	$n\sqrt{2}$	6

Simplify. Assume n represents a positve number.

$n^3 \cdot n^4$

$n^{-2} \cdot n^4$

$(n^3)^2$

$\dfrac{n^7}{n^3}$

$(3n^3)(-2n^3)$

$n^2 \cdot n^{-3}$

$\dfrac{2n^4}{4n^{-2}}$

$(2n^2 \cdot 3n^{-5})^2$

$(3n^2)^{-1}$

$\left(\dfrac{n^{-1}}{n^3}\right)^{-2}$

$(-n^3 \cdot n^{-1} \cdot -n^{-1})^{-2}$

$\sqrt{81}$

$-\sqrt[3]{125}$

$\sqrt[4]{16}$

$\sqrt[3]{-27}$

$\sqrt{2n^2}$

$\sqrt[3]{64n^6}$

$-\sqrt[4]{81n^2}$

$\sqrt[3]{-8n^6}$

Name: _____

	A	B	C	D	E	F	
1	$\frac{1}{27}$	$\frac{1}{8}$	$-\frac{1}{243}$	$54\sqrt{2}$	$\frac{1}{16}$	$\frac{1}{8}$	1
2	$24\sqrt{3}$	81	25	16	$24\sqrt{3}$	81	2
3	$2\sqrt[4]{2}$	$2\sqrt[3]{4}$	64	$2\sqrt{3}$	$2\sqrt[4]{2}$	$2\sqrt[3]{4}$	3
4	9	-2	$2\sqrt{3}$	64	9	-2	4
5	$\frac{1}{16}$	$\frac{1}{8}$	$-\frac{1}{243}$	$54\sqrt{2}$	$\frac{1}{27}$	$\frac{1}{8}$	5
6	$24\sqrt{3}$	81	25	16	$24\sqrt{3}$	81	6

Simplify without fractional exponents or radical symbols.

 $27^{\frac{2}{3}}$

 $16^{\frac{3}{2}}$

 $8^{\frac{4}{3}}$

 $9^{-\frac{3}{2}}$

 $16^{-\frac{3}{4}}$

 $(-27)^{-\frac{5}{3}}$

Simplify without a fractional exponent.

 $12^{\frac{1}{2}}$

 $(-27)^{\frac{4}{3}}$

 $18^{\frac{3}{2}}$

 $(-8)^{-\frac{4}{3}}$

 $12^{\frac{3}{2}}$

 $(-125)^{\frac{2}{3}}$

 $32^{\frac{1}{2}} \cdot 32^{-\frac{1}{4}}$

 $2^{\frac{4}{3}} \cdot 2^{\frac{1}{3}}$

 $\dfrac{(-8)^{\frac{5}{3}}}{(-8)^{\frac{4}{3}}}$

Activity 12

Name: _____

	A	B	C	D	E	F	
1	$3x\sqrt[3]{5x^2}$	$\sqrt[3]{30}$	$x\sqrt{6}$	$2x\sqrt[4]{2}$	$\sqrt{10}$	$\sqrt[4]{125}$	1
2	$\sqrt{10x}$	$\sqrt[12]{128}$	$2\sqrt[3]{2}$	3	$3x$	17	2
3	$\sqrt{35}$	$\sqrt[12]{128}$	$4\sqrt{3}$	$\sqrt{35}$	$\sqrt[12]{128}$	$4\sqrt{3}$	3
4	$2y\sqrt[4]{3x^3y}$	$\sqrt[3]{30}$	$4x$	$2y\sqrt[4]{3x^3y}$	$\sqrt[3]{30}$	$4x$	4
5	$2x\sqrt[4]{2}$	$\sqrt{10}$	$\sqrt[4]{125}$	$3x\sqrt[3]{5x^2}$	$\sqrt[3]{30}$	$x\sqrt{6}$	5
6	3	$3x$	17	$\sqrt{10x}$	$\sqrt[12]{128}$	$2\sqrt[3]{2}$	6

Simplify. Assume n represents a positive number.

$\sqrt{17} \cdot \sqrt{17}$

$\sqrt{5} \cdot \sqrt{7}$

$\sqrt{\dfrac{2}{3}} \cdot \sqrt{15}$

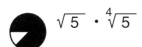 $\sqrt[3]{3} \cdot \sqrt[3]{9}$

$\sqrt[3]{4} \cdot \sqrt[3]{4}$

$\sqrt{2x} \cdot \sqrt{8x}$

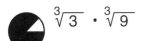 $\sqrt[3]{3x^2} \cdot \sqrt[3]{9x}$

$\sqrt[3]{3} \cdot \sqrt[3]{5} \cdot \sqrt[3]{2}$

$\sqrt[4]{2x} \cdot \sqrt[4]{8x^2} \cdot \sqrt[4]{2x}$

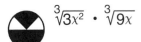 $\sqrt{48}$

$\sqrt[3]{135x^5}$

$\sqrt[4]{48x^3y^5}$

Write as one radical in simplest form.

 $\sqrt{5} \cdot \sqrt[4]{5}$  $\sqrt[3]{2} \cdot \sqrt[4]{2}$ $\sqrt{10} \cdot \sqrt[4]{x^2}$ $\sqrt{3x} \cdot \sqrt[8]{16x^4}$

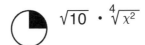

Algebra II Topics by Design / Russell F. Jacobs www.tessellations.com © 2007 Jacobs Publishing Company

Name:_____

	A	B	C	D	E	F	
1	$\sqrt[3]{2x}$	$8\sqrt[3]{3}$	$7\sqrt{3}$	$7\sqrt{3}$	$\sqrt[3]{2x}$	$5x\sqrt{xy}$
2	$5\sqrt{6}$	$5\sqrt{2}$	$3\sqrt{2}$	$-x\sqrt[3]{y}$	$7\sqrt[3]{2}$	$\sqrt[3]{5}$
3	$-\sqrt{3}$	$4\sqrt{3}$	$8\sqrt{2}$	$3\sqrt{3}$	$-\sqrt{3}$	$4\sqrt{3}$
4	$4\sqrt{3}$	$-\sqrt{3}$	$3\sqrt{2}$	$-x\sqrt[3]{y}$	$4\sqrt{3}$	$-\sqrt{3}$
5	$\sqrt[3]{2x}$	$5x\sqrt{xy}$	$8\sqrt{2}$	$3\sqrt{3}$	$\sqrt[3]{2x}$	$8\sqrt[3]{3}$
6	$7\sqrt[3]{2}$	$\sqrt[3]{5}$	$\sqrt{7}$	$\sqrt{7}$	$5\sqrt{6}$	$5\sqrt{2}$

Simplify each sum or difference. Assume x and y represent positive numbers.

 $5\sqrt{2} \ - \ 2\sqrt{2}$

 $4\sqrt{3} \ + \ 3\sqrt{3}$

 $\sqrt{12} \ - \ 3\sqrt{3}$

 $\sqrt{18} + \sqrt{8}$

 $\sqrt{28} - \sqrt{7}$

$\sqrt{24} + \sqrt{54}$

 $\sqrt{50} + \sqrt{72} - \sqrt{18}$

$\sqrt{27} - \sqrt{48} + \sqrt{75}$

$\sqrt{108} + \sqrt{147} - \sqrt{300}$

$\sqrt[3]{54} + \sqrt[3]{128}$

$\sqrt[3]{81} + \sqrt[3]{375}$

$\sqrt[3]{135} - \sqrt[3]{40}$

 $2\sqrt{x^3 y} + 3x\sqrt{xy}$

$2\sqrt[3]{x^3 y} - 3x\sqrt[3]{y}$

$\sqrt[3]{128x} - \sqrt[3]{54x}$

Activity 14

Name:_____

	A	B	C	D	E	F	
1	(x^2+4y)	$4xy$	x^4	$(2x+y+3)$	(x^2-4y)	$(x-2)$	1
2	$2xy$	$(x-1)$	$(3x+4)$	$2x$	$(xy-2)$	$(x+5)$	2
3	7	(x^2+y^2)	$(4x+5y)$	$(4x-5y)$	$(2x-1)$	$(x+y)(x-y)$	3
4	x	$2(2x+3)$	$(2x+5)$	$9x$	$(3x-10)$	$(2x-3)$	4
5	(x^2-4y)	$(2x+3y)$	$(9+y^4)$	$(2x-y-3)$	(x^2+4y)	$(x-2)$	5
6	$(xy-2)$	$(x+5)$	$(3x-4)$	$2x$	$2xy$	$(x-1)$	6

Factor completely. Look for *all* the factors on the grid.

 $3x^2 - 10x$

$2x^2 + 6x$

$18x^2 + 45x$

 $8x^2y + 12xy^2$

$9x^2 - 16$

$2x^2y^2 - 4xy$

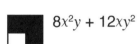 $16x^2 - 25y^2$

$8x^2 - 18$

$3(x-2) + x(x-2)$

$x^4 - 16y^2$

$x^4 - y^4$

$9x^4 + x^4y^4$

$(x+2)^2 - 9$

$4x^2 - (y+3)^2$

$(x+3)^2 - (x-4)^2$

Activity 15

Name: _____

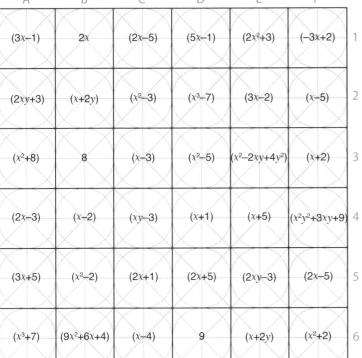

	A	B	C	D	E	F	
	$(3x-1)$	$2x$	$(2x-5)$	$(5x-1)$	$(2x^2+3)$	$(-3x+2)$	1
	$(2xy+3)$	$(x+2y)$	(x^2-3)	(x^3-7)	$(3x-2)$	$(x-5)$	2
	(x^2+8)	8	$(x-3)$	(x^2-5)	$(x^2-2xy+4y^2)$	$(x+2)$	3
	$(2x-3)$	$(x-2)$	$(xy-3)$	$(x+1)$	$(x+5)$	$(x^2y^2+3xy+9)$	4
	$(3x+5)$	(x^2-2)	$(2x+1)$	$(2x+5)$	$(2xy-3)$	$(2x-5)$	5
	(x^3+7)	$(9x^2+6x+4)$	$(x-4)$	9	$(x+2y)$	(x^2+2)	6

Factor completely. Look for *all* the factors on the grid.

 $x^2 - x - 6$

 $x^2 - 9x + 20$

 $2x^2 - x - 3$

 $-6x^2 + x + 2$

 $6x^2 + 13x - 5$

 $x^4 - x^2 - 6$

 $2x^4 - x^2 - 6$

 $4x^2 - 20x + 25$

 $2x^3 + 6x^2 - 20x$

 $27x^3 - 8$

 $15x^2 + 22x - 5$

 $8x^3 + 64y^3$

 $36x^2y^2 - 81$

 $x^4 + 3x^2 - 40$

$x^6 - 49$

$x^3y^3 - 27$

Name:_____

	A	B	C	D	E	F	
	$\left\{\dfrac{3}{2}, \dfrac{2}{3}\right\}$	$\{3, -3\}$	$\{3, 5\}$	$\{1, 2\}$	$\{-2, -4\}$	$\left\{-\dfrac{3}{2}, -5\right\}$	1
	$\{2\}$	$\left\{-\dfrac{3}{5}, \dfrac{1}{2}\right\}$	$\left\{\dfrac{3}{4}, -\dfrac{1}{2}\right\}$	$\left\{\dfrac{3}{2}, -2\right\}$	$\{0, 10\}$	$\{2\}$	2
	$\{3, 4\}$	$\{-2, -4\}$	$\left\{\dfrac{4}{5}, \dfrac{1}{3}\right\}$	$\left\{-\dfrac{1}{3}, 5\right\}$	$\{3, -3\}$	$\{5, -3\}$	3
	$\{1, 2\}$	$\{-2, -4\}$	$\left\{-\dfrac{3}{2}, -5\right\}$	$\left\{\dfrac{3}{2}, \dfrac{2}{3}\right\}$	$\{3, -3\}$	$\{3, 5\}$	4
	$\left\{\dfrac{3}{4}, -\dfrac{1}{2}\right\}$	$\{0, 10\}$	$\{2\}$	$\{2\}$	$\left\{-\dfrac{3}{5}, \dfrac{1}{2}\right\}$	$\left\{\dfrac{3}{2}, -2\right\}$	5
	$\left\{-\dfrac{1}{3}, 5\right\}$	$\{3, -3\}$	$\{5, -3\}$	$\{3, 4\}$	$\{-2, -4\}$	$\left\{\dfrac{4}{5}, \dfrac{1}{3}\right\}$	6

Find each solution set by factoring.

Example: $x^2 - x - 6 = 0$
$(x + 2)(x - 3) = 0$
$\{-2, 3\}$

 $x^2 - 3x + 2 = 0$

 $x^2 - 2x - 15 = 0$

$x^2 + 6x + 8 = 0$

 $x^2 - 7x + 12 = 0$

 $10x^2 + x - 3 = 0$

 $8x^2 - 2x - 3 = 0$

 $2x^2 + 13x + 15 = 0$

 $x^2 - 8x + 15 = 0$

$x^2 - 10x = 0$

$2x^2 + x - 6 = 0$

$3x^2 - 14x - 5 = 0$

$x^2 - 9 = 0$

$x^2 - 4x + 4 = 0$

$-6x^2 + 13x - 6 = 0$

$15x^2 - 17x + 4 = 0$

Name:_____

Refer to the quadratic formula for the following excercises.

Quadratic Formula: $x = \dfrac{-b \pm \sqrt{b^2 - 4ac}}{2a}$

	A	B	C	D	E	F	
1	$\left\{\frac{1}{4}, -\frac{3}{2}\right\}$	40	$\left\{\sqrt{5}, -\sqrt{5}\right\}$	28	$\left\{\frac{1}{4}, -\frac{3}{2}\right\}$	$\left\{-\frac{1}{3}, 2\right\}$
2	$\left\{-\frac{1}{3}, 2\right\}$	$\left\{\frac{1}{4}, -\frac{3}{2}\right\}$	20	24	40	$\left\{\frac{1}{4}, -\frac{3}{2}\right\}$
3	$\left\{\frac{1}{2}, 1\right\}$	$\left\{1, -\frac{5}{3}\right\}$	89	8	$\left\{\frac{1}{2}, 1\right\}$	$\left\{1, -\frac{5}{3}\right\}$
4	$\left\{1, -\frac{5}{3}\right\}$	$\left\{\frac{1}{2}, 1\right\}$	25	$\left\{2\sqrt{3}, -2\sqrt{3}\right\}$	$\left\{1, -\frac{5}{3}\right\}$	$\left\{\frac{1}{2}, 1\right\}$
5	$\left\{\frac{1}{4}, -\frac{3}{2}\right\}$	$\left\{-\frac{1}{3}, 2\right\}$	$\left\{\sqrt{5}, -\sqrt{5}\right\}$	28	$\left\{\frac{1}{4}, -\frac{3}{2}\right\}$	40
6	40	$\left\{\frac{1}{4}, -\frac{3}{2}\right\}$	20	24	$\left\{-\frac{1}{3}, 2\right\}$	$\left\{\frac{1}{4}, -\frac{3}{2}\right\}$

Find the quadratic discriminant.

$x^2 - 2x - 1 = 0$

$2x^2 - 8x + 5 = 0$

$-3x^2 = x - 2$

$x^2 = 6 - 4x$

$4x^2 + 1 = 6x$

$5x^2 - 4 = 3x$

$6x - 1 = 2x^2$

Find the solution set of each equation.

$2x^2 - 3x + 1 = 0$

$3x^2 - 5x - 2 = 0$

$8x^2 = 3 - 10x$

$-5 = -x^2$

$3x^2 + 2x = 5$

$x^2 - 12 = 0$

Name:_____

	A	B	C	D	E	F	
	$2x^2-3x-2=0$	$x^2-3x-10=0$	$8x^2-2x-1=0$	$x^2-x-6=0$	$x^2-5x+4=0$	$4x^2-17x+4=0$	1
	$8x^2+10x-3=0$	$2x^2-7x-4=0$	$3x^2+7x+2=0$	$3x^2+7x+2=0$	$x^2+4x+3=0$	$8x^2+10x-3=0$	2
	$10x^2-7x+1=0$	$3x^2+4x+1=0$	$2x^2+7x-4=0$	$4x^2+4x-3=0$	$3x^2-5x-2=0$	$2x^2+7x+6=0$	3
	$x^2-x-6=0$	$3x^2+4x+1=0$	$4x^2-17x+4=0$	$2x^2-3x-2=0$	$x^2-3x-10=0$	$8x^2-2x-1=0$	4
	$3x^2+7x+2=0$	$x^2+4x+3=0$	$8x^2+10x-3=0$	$8x^2+10x-3=0$	$2x^2-7x-4=0$	$3x^2+7x+2=0$	5
	$4x^2+4x-3=0$	$3x^2-5x-2=0$	$2x^2+7x+6=0$	$10x^2-7x+1=0$	$x^2-5x+4=0$	$2x^2+7x-4=0$	6

Write a quadratic equation with *integral* coefficients for each solution set.

Example:

$\left\{\dfrac{1}{2},\ 3\right\}$

$(x-\dfrac{1}{2})(x-3)=0$

$x^2-\dfrac{7}{2}x+\dfrac{3}{2}=0$

so

$2x^2-7x+3=0$

$\{5,-2\}$

$\left\{\dfrac{1}{4},\ 4\right\}$

$\left\{-\dfrac{3}{2},\ \dfrac{1}{4}\right\}$

$\{-1.5,-2\}$

$\{-3,-1\}$

$\left\{-\dfrac{1}{3},-2\right\}$

$\left\{-\dfrac{1}{3},\ 2\right\}$

$\{0.2, 0.5\}$

$\{0.5,-4\}$

$\{3,-2\}$

$\{4, 1\}$

$\left\{-\dfrac{1}{2},\ 2\right\}$

$\left\{4,-\dfrac{1}{2}\right\}$

$\left\{-1,-\dfrac{1}{3}\right\}$

$\left\{\dfrac{1}{2},-\dfrac{1}{4}\right\}$

$\left\{\dfrac{1}{2},-\dfrac{3}{2}\right\}$

Activity 19

Name: _____

	A	B	C	D	E	F	
1	−2	8	−12	(0,−2)	2	8	
2	−8	2	−14	0	−8	−2	
3	11	(−4,4)	3	−7	11	(−4,4)	
4	11	(−4,4)	−25	−6	11	(−4,4)	
5	2	−8	−12	(0,−2)	−2	−8	
6	8	−2	−14	0	8	2	

Functions are often described by equations, like $y = 2x − 5$ or $f(x) = 2x − 5$. $f(2)$ is the value of y if x equals 2. Thus, $(2, −1)$ is an ordered pair of $f(x) = 2x − 5$.

Find the indicated value of each function.

$f(x) = x^2 − 2x + 3$
$f(4) =$

$g(x) = −2x^2 − 5$
$g(−1) =$

$h(x) = −x^2 + 3x − 2$
$h(−2) =$

$f(x) = 4x^2 − x − 6$
$f(2) =$

$g(x) = −2x^2 + 7x − 3$
$g(−2) =$

$h(x) = 2x^2 − x − 5$
$h(−1) =$

$f(x) = 2 (x − 3)^2$
$f(2) =$

$g(x) = −3 (x − 1)^2 − 5$
$g(0) =$

$h(x) = − (x + 2)^2 + 3$
$h(1) =$

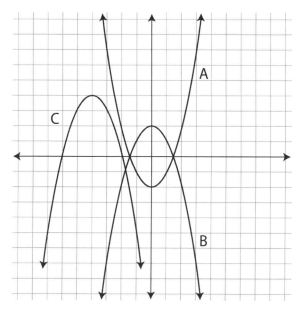

Graphs A, B, and C to the left represent functions.

What are the coordinates of the lowest point of A?

What are the coordinates of the highest point of C?

In graph A, what positive x value corresponds to a y value of 7?

If the equation of graph C is $y = −(x + 4)^2 + 4$, what is the value of y if x equals −2?

If the equation of graph B is $y = −x^2 + 2$, what is the value of y if x equals 4?

Activity 20

Name: _____

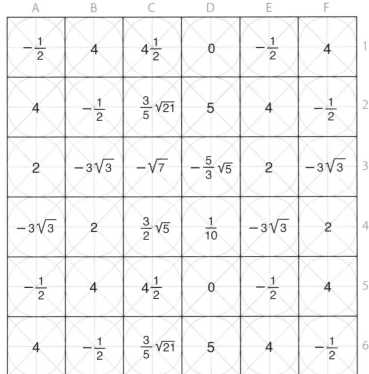

	A	B	C	D	E	F	
1	$-\frac{1}{2}$	4	$4\frac{1}{2}$	0	$-\frac{1}{2}$	4	1
2	4	$-\frac{1}{2}$	$\frac{3}{5}\sqrt{21}$	5	4	$-\frac{1}{2}$	2
3	2	$-3\sqrt{3}$	$-\sqrt{7}$	$-\frac{5}{3}\sqrt{5}$	2	$-3\sqrt{3}$	3
4	$-3\sqrt{3}$	2	$\frac{3}{2}\sqrt{5}$	$\frac{1}{10}$	$-3\sqrt{3}$	2	4
5	$-\frac{1}{2}$	4	$4\frac{1}{2}$	0	$-\frac{1}{2}$	4	5
6	4	$-\frac{1}{2}$	$\frac{3}{5}\sqrt{21}$	5	4	$-\frac{1}{2}$	6

 In graph A below, what is the value of y if x equals 0?

 In graph B below, what is the value of x if y equals 2?

 Graph B has the equation $x = \frac{1}{2}y^2$. What is the value of x if y equals 3?

 The equation for graph E is $\frac{x^2}{4} - \frac{y^2}{9} = 1$. What is the positive value of y if x equals 3?

 In graph E, what is the negative value of y if x equals 4?

 The equation of graph F is $y = -\frac{1}{x}$. What is the value of y if x equals -10?

 In graph C, what is the positive value of y if x equals 0?

 In graph D, what is the absolute value of x if y equals 2?

 The equation for graph D is $\frac{x^2}{9} + \frac{y^2}{25} = 1$. What is the negative value of y if x equals 2?

 In graph F, what is the value of y if x equals 2?

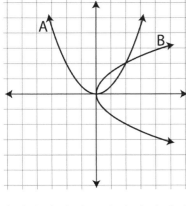

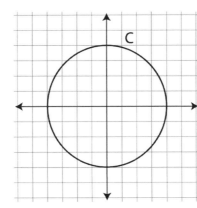

 The equation of graph C is $x^2 + y^2 = 16$. What is the negative value of y if x equals 3?

 In graph D, what is the positive value of y if x equals 0?

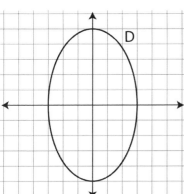

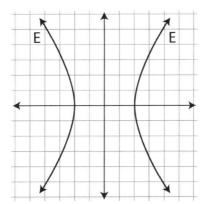

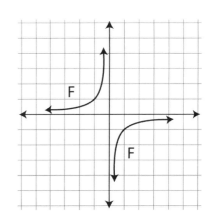

Activity 21

Name:_____

Simplify each expression.

	A	B	C	D	E	F	
1	$\dfrac{x-4}{2}$	$\dfrac{3}{x-1}$	$\dfrac{2}{x}$	$\dfrac{x}{2(x+1)}$	$\dfrac{x-4}{2}$	$\dfrac{3}{x-1}$	1
2	$\dfrac{x-3}{x+3}$	$\dfrac{1}{2}$	$\dfrac{x+4}{2(x-4)}$	$\dfrac{x}{2}$	$\dfrac{1}{x+5}$	$\dfrac{1}{2}$	2
3	$\dfrac{1}{x+3}$	$\dfrac{x+3}{x+1}$	$\dfrac{x+5}{x-5}$	$\dfrac{x+1}{2(x-1)}$	$\dfrac{1}{x+3}$	$\dfrac{x+3}{x+1}$	3
4	$\dfrac{1}{x+3}$	$\dfrac{x+3}{x+1}$	$\dfrac{x-2}{x-1}$	$-\dfrac{1}{x+1}$	$\dfrac{1}{x+3}$	$\dfrac{x+3}{x+1}$	4
5	$\dfrac{x-4}{2}$	$\dfrac{3}{x-1}$	$\dfrac{2}{x}$	$\dfrac{x}{2(x+1)}$	$\dfrac{x-4}{2}$	$\dfrac{3}{x-1}$	5
6	$\dfrac{1}{x+5}$	$\dfrac{1}{2}$	$\dfrac{x+4}{2(x-4)}$	$\dfrac{x}{2}$	$\dfrac{x-3}{x+3}$	$\dfrac{1}{2}$	6

 $\dfrac{x-2}{x^2+x-6}$

 $\dfrac{x^2+4x}{2x^2-8x}$

$\dfrac{x^2-4}{x^2+x-2}$

 $\dfrac{3x^2+3x}{x^3-x}$

$\dfrac{x^2+x-6}{x^2-x-2}$

$\dfrac{2x^2-4x-6}{4x^2-16x+12}$

 $\dfrac{x^2-16}{2x+8}$

$\dfrac{2x^2-10x}{4x^2-16x-20}$

$\dfrac{x^2-6x+9}{x^2-9}$

$\dfrac{x^2+6x+5}{x^2-4x-5}$

$\dfrac{5x^2+x}{10x^2+2x}$

$\dfrac{8x-4}{4x^2-2x}$

 $\dfrac{x-5}{x^2-25}$

$\dfrac{4-x}{x^2-3x-4}$

 $\dfrac{6x-5x^2+x^3}{2x^2-10x+12}$

Algebra II Topics by Design / Russell F. Jacobs   www.tessellations.com   © 2007 Jacobs Publishing Company

Name:_____

Simplify each product or quotient.

	A	B	C	D	E	F	
	$\dfrac{3(x-3)}{2}$	$\dfrac{3x}{4}$	3	$\dfrac{x+1}{x}$	$\dfrac{x}{3}$	$\dfrac{x}{2}$	1
	$\dfrac{x+2}{x+4}$	$\dfrac{x+2}{4x}$	$\dfrac{2(x-2)}{x}$	$\dfrac{2(x-2)}{x}$	$\dfrac{x+2}{4x}$	$\dfrac{x+2}{x+4}$	2
	$\dfrac{(x+1)(x-2)}{5}$	$\dfrac{x}{3}$	$\dfrac{1}{3}$	$\dfrac{1}{2}$	$\dfrac{3x}{4}$	$\dfrac{2x}{3(x-5)}$	3
	$2x(x+1)$	$\dfrac{x}{3}$	$\dfrac{x}{2}$	$\dfrac{3(x-3)}{2}$	$\dfrac{3x}{4}$	$x-2$	4
	$\dfrac{2(x-2)}{x}$	$\dfrac{x+2}{4x}$	$\dfrac{x+2}{x+4}$	$\dfrac{x+2}{x+4}$	$\dfrac{x+2}{4x}$	$\dfrac{2(x-2)}{x}$	5
	$\dfrac{1}{2}$	$\dfrac{3x}{4}$	$\dfrac{2x}{3(x-5)}$	$\dfrac{(x+1)(x-2)}{5}$	$\dfrac{x}{3}$	$\dfrac{1}{3}$	6

$\dfrac{2}{x-2} \cdot \dfrac{x^2-2x}{4}$

$\dfrac{x+1}{2} \cdot \dfrac{x-1}{x^2-1}$

$\dfrac{1}{x-3} \cdot \dfrac{x^2-x-6}{x+4}$

$\dfrac{2x^2-4x}{6x} \cdot \dfrac{x+2}{x^2-4}$

$\dfrac{3x}{x+3} \cdot \dfrac{x^2-9}{4x-12}$

$\dfrac{x^2-3x-4}{5} \cdot \dfrac{x^2-4}{x^2-2x-8}$

$\dfrac{2x^3+8x^2}{x-4} \cdot \dfrac{x^2-3x-4}{x^2+4x}$

$\dfrac{x^2-4}{x+1} \cdot \dfrac{x+1}{x+2}$

$\dfrac{2x^2-3x}{2x-10} \cdot \dfrac{4}{3(2x-3)}$

$\dfrac{2x}{3x-1} \div \dfrac{6x}{3x^2-x}$

$\dfrac{2x+2}{x+2} \div \dfrac{x^2+x}{x^2-4}$

$\dfrac{x^2-9}{2x^2+8x} \div \dfrac{x+3}{3x^2+12x}$

$\dfrac{2x^2-x-1}{2x^2-3x+1} \div \dfrac{2x^2+x}{2x^2+x-1}$

$\dfrac{3x^2+6x}{x^2+x} \div \dfrac{x^2+x-2}{x^2-1}$

$\dfrac{x^2-4}{2x^2+2x} \div \dfrac{2x-4}{x+1}$

Name:_____

	A	B	C	D	E	F	
1	$\dfrac{x+13}{6x}$	$\dfrac{x+11}{2x}$	$\dfrac{-4x+10}{(x+2)(x-2)}$	$\dfrac{-4x+10}{(x+2)(x-2)}$	$\dfrac{x+13}{6x}$	$\dfrac{x+11}{2x}$	1
2	$\dfrac{5}{x-3}$	$\dfrac{2x-1}{x-1}$	$\dfrac{3x^2+x+5}{5(x+2)}$	$\dfrac{-3x+17}{2x(x-1)}$	$\dfrac{5}{x-3}$	$\dfrac{2x-1}{x-1}$	2
3	$\dfrac{x+29}{4(x-3)}$	$\dfrac{2x^2+17x}{5(x+1)}$	$\dfrac{x-1}{x+1}$	$\dfrac{2x^2-4}{(x+1)(x+2)}$	$\dfrac{x+29}{4(x-3)}$	$\dfrac{2x^2+17x}{5(x+1)}$	3
4	$\dfrac{2x^2+17x}{5(x+1)}$	$\dfrac{x+29}{4(x-3)}$	$\dfrac{3x^2+x+5}{5(x+2)}$	$\dfrac{-3x+17}{2x(x-1)}$	$\dfrac{2x^2+17x}{5(x+1)}$	$\dfrac{x+29}{4(x-3)}$	4
5	$\dfrac{x+13}{6x}$	$\dfrac{x+11}{2x}$	$\dfrac{x-1}{x+1}$	$\dfrac{2x^2-4}{(x+1)(x+2)}$	$\dfrac{x+13}{6x}$	$\dfrac{x+11}{2x}$	5
6	$\dfrac{5}{x-3}$	$\dfrac{2x-1}{x-1}$	$\dfrac{x^2-2x+4}{3(x-2)}$	$\dfrac{x^2-2x+4}{3(x-2)}$	$\dfrac{5}{x-3}$	$\dfrac{2x-1}{x-1}$	6

Simplify each sum or difference.

$\dfrac{2}{x-3} + \dfrac{3}{x-3}$

$\dfrac{x+1}{x-1} + \dfrac{x-2}{x-1}$

$\dfrac{2x}{5} + \dfrac{3x}{x+1}$

$\dfrac{x+5}{x-3} - \dfrac{3}{4}$

$\dfrac{x}{x^2-4} - \dfrac{5}{x+2}$

$\dfrac{4}{3x-6} + \dfrac{x}{3}$

$\dfrac{3x}{5} - \dfrac{x-1}{x+2}$

$\dfrac{x+6}{x^2-x} - \dfrac{5}{2x}$

$\dfrac{x-1}{x+1} + \dfrac{x-2}{x+2}$

$\dfrac{x+1}{2x} - \dfrac{x-5}{3x}$

$\dfrac{x}{x+2} - \dfrac{2}{x^2+3x+2}$

$\dfrac{x+4}{x} - \dfrac{x-3}{2x}$

Activity 24

Name:_____

	A	B	C	D	E	F	
1	(0,–2)	(–4,2)	(5,4)	(5,–2)	(0,–2)	(–4,2)
2	($\frac{1}{2}$,0)	(–3,–5)	(3,3)	(–4,3)	($\frac{1}{2}$,0)	(3,2)
3	(5,–5)	(–2,3)	(–4,–2)	(–6,3)	(5,–5)	(–2,3)
4	(3,–5)	(–3,–4)	(–4,–2)	(–6,3)	(3,–5)	(–3,–4)
5	(0,–2)	(–4,2)	(5,4)	(5,–2)	(0,–2)	(–4,2)
6	($\frac{1}{2}$,0)	(3,2)	(3,3)	(–4,3)	($\frac{1}{2}$,0)	(–3,–5)

Solve each system of linear equations.

$\begin{cases} x + y - 1 = 0 \\ -2x + y - 7 = 0 \end{cases}$

$\begin{cases} -x + 4y + 4 = 0 \\ x + 4y + 12 = 0 \end{cases}$

$\begin{cases} x - 3y - 12 = 0 \\ x + y + 8 = 0 \end{cases}$

$\begin{cases} x - 3y - 18 = 0 \\ 7x + 3y - 6 = 0 \end{cases}$

$\begin{cases} x + y - 9 = 0 \\ 2x - 5y + 10 = 0 \end{cases}$

$\begin{cases} x - 4y - 8 = 0 \\ 5x + 2y + 4 = 0 \end{cases}$

$\begin{cases} 3x - 2y - 5 = 0 \\ x + y - 5 = 0 \end{cases}$

$\begin{cases} 5x - 6y + 48 = 0 \\ x - 3y + 15 = 0 \end{cases}$

$\begin{cases} x + y + 2 = 0 \\ 2x + y + 6 = 0 \end{cases}$

$\begin{cases} 3x - 4y + 24 = 0 \\ x + 4y - 8 = 0 \end{cases}$

$\begin{cases} 2x - 5y - 35 = 0 \\ 6x + 5y - 5 = 0 \end{cases}$

$\begin{cases} 4x - 3y - 2 = 0 \\ 2x + y - 1 = 0 \end{cases}$

$\begin{cases} 2x - 3y + 3 = 0 \\ -x - y + 6 = 0 \end{cases}$

$\begin{cases} 2x - 3y - 6 = 0 \\ 2x - y + 2 = 0 \end{cases}$

$\begin{cases} 2x - 5y - 20 = 0 \\ 6x + 5y - 20 = 0 \end{cases}$

Name:_____

Write an inequality that describes each graph.

	A	B	C	D	E	F	
	$y \leq -\frac{4}{7}x+1$	$y < \frac{4}{3}x+4$	$y \geq 2x-4$	$y \geq \frac{1}{2}x-1$	$y < \frac{4}{3}x+4$	$y \leq \frac{1}{4}x$	1
	$y < -2x-2$	$y \leq \frac{1}{4}x$	$y \geq \frac{1}{2}x-1$	$y \geq 2x-4$	$y \leq -\frac{4}{7}x+1$	$y < -2x-2$	2
	$y > -x+3$	$y \geq -\frac{3}{4}x-3$	$y \leq -x+3$	$y \leq \frac{1}{2}x+2$	$y > -x+3$	$y \geq -\frac{3}{4}x-3$	3
	$y > -\frac{5}{2}x+5$	$y < 3$	$y < x-1$	$x \leq 3$	$y > -\frac{5}{2}x+5$	$y < 3$	4
	$y < \frac{4}{3}x+4$	$y \leq \frac{1}{4}x$	$y \geq 2x-4$	$y \geq \frac{1}{2}x-1$	$y \leq -\frac{4}{7}x+1$	$y < \frac{4}{3}x+4$	5
	$y \leq -\frac{4}{7}x+1$	$y < -2x-2$	$y \geq \frac{1}{2}x-1$	$y \geq 2x-4$	$y < -2x-2$	$y \leq \frac{1}{4}x$	6

Activity 26

Name: _____

	A	B	C	D	E	F	
	$y \geq \frac{1}{2}x - 2$	$y < -2x + 2$	$y \geq \frac{3}{2}x - 3$	$y > x - 2$	$y \geq \frac{1}{2}x - 2$	$y < -2x + 2$	1
	$y \leq -2x - 1$	$y > -2x - 1$	$y > x - 2$	$y \geq \frac{3}{2}x - 3$	$x > -3$	$y > -2x - 1$	2
	$y < -3x + 1$	$y < 2x + 1$	$y > x - 2$	$y \geq \frac{3}{2}x - 3$	$y < 2x + 1$	$y < 2x$	3
	$y \leq -x + 3$	$y < 2x$	$y \geq \frac{3}{2}x - 3$	$y > x - 2$	$y < -3x + 1$	$y \leq -x + 3$	4
	$y \geq x - 4$	$y < -2x + 2$	$y \geq \frac{3}{2}x - 3$	$y > x - 2$	$y \geq x - 4$	$y < -2x + 2$	5
	$y \leq -2x - 1$	$y > 2x - 1$	$y > x - 2$	$y \geq \frac{3}{2}x - 3$	$x > -3$	$y > -2x - 1$	6

Each grid has the graph of a system of linear inequalities. One inequality is given. Find the other.

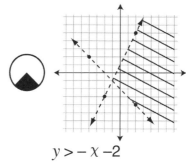
$y > -x - 2$

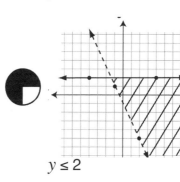
$y < -2x + 1$

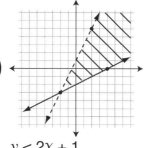
$y \geq x - 3$

$y \leq 3x - 1$

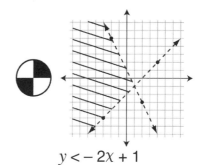
$y \leq 2$

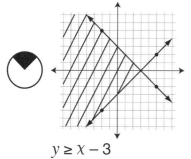
$y < 2x + 1$

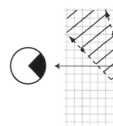
$y > -x - 3$

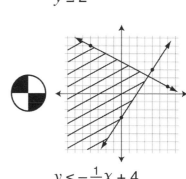
$y \leq -\frac{1}{2}x + 4$

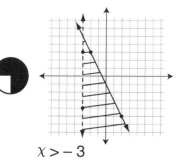
$x > -3$

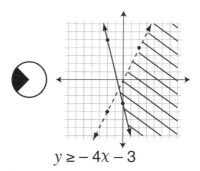
$y \geq -4x - 3$

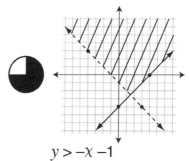
$y > -x - 1$

Name:_____

Write each complex number in the form $a + bi$.

	A	B	C	D	E	F	
	$6+i$	$-1+i$	$6+(-4i)$	$-5+4i$	$6+i$	$0+(-i)$	1
	$0+(-i)$	$6+i$	$-2+2i$	$-6+(-4i)$	$-1+i$	$6+i$	2
	$0+2i$	$11+3i$	$3+0i$	$-3+2i$	$0+2i$	$11+3i$	3
	$-1+(-4i)$	$-1+5i$	$2+(-2i)$	$3+i$	$-1+(-4i)$	$-1+5i$	4
	$6+i$	$0+(-i)$	$6+(-4i)$	$-5+4i$	$6+i$	$-1+i$	5
	$-1+i$	$6+i$	$-2+2i$	$-6+(-4i)$	$0+(-i)$	$6+i$	6

$-5 + 2i + 3$

$3i - 1 + 2i$

$-5i - 6 + i$

$4 - i + 2i - 1$

$-6 + 2i + 5 - i$

$-i + 5 + 1 - 3i$

$3i - i^2 + 10$

$4 - i - 2 + i^3$

$(3 - i) - (4 + 3i)$

$(5 + i) - (-2 - i^2)$

$(-12 + 3i) - (-7 - i)$

$(3 - i) - (2 + i^4)$

$(4 + i^2) + (-1 - i^2)$

$(1 - i^3) - (1 + i^3)$

$(1 - 2i) + (3 + i) - (7 - 3i)$

Name:_____

Write the complex number represented by each labeled point in the complex plane.

	A	B	C	D	E	F	
	5+0*i*	−4−5*i*	−3+6*i*	2+4*i*	6−8*i*	−4−5*i*	1
	0−4*i*	−8+10*i*	8+5*i*	0+6*i*	7−5*i*	−8+10*i*	2
	3−3*i*	−8−2*i*	−6+*i*	4+8*i*	3−3*i*	−8−2*i*	3
	3−3*i*	−8−2*i*	−6+4*i*	8+10*i*	3−3*i*	−8−2*i*	4
	5+0*i*	−4−5*i*	−3+6*i*	2+4*i*	6−8*i*	−4−5*i*	5
	0−4*i*	−8+10*i*	8+5*i*	0+6*i*	7−5*i*	−8+10*i*	6

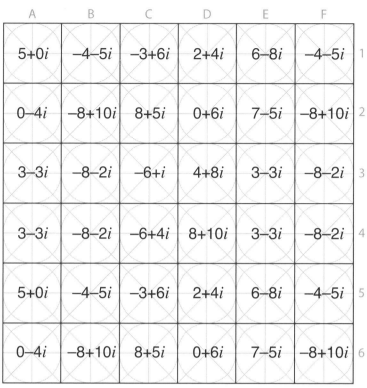

 A

 B

 C

 D

 E

 F

 G

 H

 I

 J

 K

 L

 M

 N

 O

 P

Activity 29

Name:_____

	A	B	C	D	E	F	
	$-10i$	$3i\sqrt{5}$	$-3\sqrt{2}$	$9+i$	$4i\sqrt{3}$	$3i\sqrt{5}$	1
	$5-i\sqrt{3}$	25	$-2i\sqrt{15}$	$-3-15i$	$5-i\sqrt{3}$	25	2
	$5+12i$	$10-5i$	$8+i$	-6	$5+12i$	$10-5i$	3
	$-7+17i$	$2i\sqrt{6}$	$8+i$	-6	$-7+17i$	$2i\sqrt{6}$	4
	$4i\sqrt{3}$	$3i\sqrt{5}$	$-3\sqrt{2}$	$9+i$	$-10i$	$3i\sqrt{5}$	5
	$5-i\sqrt{3}$	25	$-2i\sqrt{15}$	$-3-15i$	$5-i\sqrt{3}$	25	6

Write the complex number for each product.

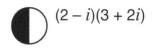 $(2 - i)(3 + 2i)$

$(4 + 3i)(1 - 2i)$

$(3 - 3i)(2 - 3i)$

$(4 + 3i)(4 - 3i)$ $(3 + 2i)(3 + 2i)$ $(1 - 3i)(3 - i)$

$(5 - i)(-2 + 3i)$ $(4 + 5i)(1 - i)$ $(\sqrt{3} + i)(\sqrt{3} - 2i)$

Simplify each radical or product of radicals.

 $\sqrt{-24}$ $\sqrt{-45}$ $\sqrt{-48}$

 $\sqrt{-3} \cdot \sqrt{-6}$ 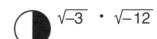 $\sqrt{-3} \cdot \sqrt{-12}$ 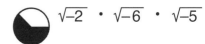 $\sqrt{-2} \cdot \sqrt{-6} \cdot \sqrt{-5}$

Algebra II Topics by Design / Russell F. Jacobs   www.tessellations.com   © 2007 Jacobs Publishing Company

Activity 30

Name:_____

	A	B	C	D	E	F	
1	$1-i$	$\frac{1}{2}+\frac{1}{2}i$	$-i\sqrt{6}$	$-\frac{5}{13}+\frac{12}{13}i$	$-\frac{6}{13}+\frac{9}{13}i$	$\frac{1}{2}+\frac{1}{2}i$	1
2	$\frac{1}{4}+\frac{7}{4}i$	$-2i$	$\frac{11}{13}+\frac{3}{13}i$	$\sqrt{2}-\frac{2}{5}i$	$\frac{1}{4}+\frac{7}{4}i$	$-2i$	2
3	i	$-\frac{1}{2}i\sqrt{2}$	$-\frac{11}{10}+\frac{17}{10}i$	$\frac{4}{5}-\frac{7}{5}i$	i	$-\frac{1}{2}i\sqrt{2}$	3
4	i	$-\frac{1}{2}i\sqrt{2}$	$-4i$	$2i$	i	$-\frac{1}{2}i\sqrt{2}$	4
5	$-\frac{6}{13}+\frac{9}{13}i$	$\frac{1}{2}+\frac{1}{2}i$	$-i\sqrt{6}$	$-\frac{5}{13}+\frac{12}{13}i$	$1-i$	$\frac{1}{2}+\frac{1}{2}i$	5
6	$\frac{1}{4}+\frac{7}{4}i$	$-2i$	$\frac{11}{13}+\frac{3}{13}i$	$\sqrt{2}-\frac{2}{5}i$	$\frac{1}{4}+\frac{7}{4}i$	$-2i$	6

Write each indicated quotient as a single complex number.

$$\frac{3-2i}{2+i}$$

$$\frac{1+3i}{2+3i}$$

$$\frac{2+i}{1-2i}$$

$$\frac{4+3i}{2-2i}$$

$$\frac{4+5i}{1-3i}$$

$$\frac{3i}{3-2i}$$

$$\frac{3+i}{4-2i}$$

$$\frac{2+3i}{2-3i}$$

$$\frac{4}{2+2i}$$

$$\frac{2}{\sqrt{-8}}$$

$$\frac{\sqrt{24}}{\sqrt{-4}}$$

$$\frac{\sqrt{32}}{\sqrt{-8}}$$

$$\frac{-\sqrt{28}}{\sqrt{-7}}$$

$$\frac{\sqrt{48}}{\sqrt{-3}}$$

$$\frac{2+i\sqrt{50}}{5i}$$

Activity 31

Name: _____

	A	B	C	D	E	F	
1	$-\dfrac{4}{5}$	$\dfrac{3}{2}$	$-\dfrac{5}{3}$	$x^2-6x+10=0$	$-\dfrac{4}{5}$	$\dfrac{3}{2}$	1
2	$\dfrac{-1 \pm i\sqrt{7}}{4}$	$1 \pm i$	$\dfrac{1 \pm i\sqrt{7}}{2}$	$\dfrac{1 \pm i\sqrt{3}}{2}$	$-\dfrac{8}{3}$	$1 \pm i$	2
3	$16x^2+8x-3=0$	$\dfrac{3 \pm i\sqrt{7}}{2}$	$-\dfrac{4}{3}$	$-\dfrac{4}{3}$	$16x^2+8x-3=0$	$\dfrac{3 \pm i\sqrt{7}}{2}$	3
4	$\dfrac{1 \pm i\sqrt{11}}{6}$	$-\dfrac{5}{2}$	$x^2-7x+12=0$	$x^2-7x+12=0$	$\dfrac{1 \pm i\sqrt{11}}{6}$	$-\dfrac{5}{2}$	4
5	$-\dfrac{4}{5}$	$\dfrac{3}{2}$	$-\dfrac{5}{3}$	$x^2-6x+10=0$	$-\dfrac{4}{5}$	$\dfrac{3}{2}$	5
6	$-\dfrac{8}{3}$	$1 \pm i$	$\dfrac{1 \pm i\sqrt{7}}{2}$	$\dfrac{1 \pm i\sqrt{3}}{2}$	$\dfrac{-1 \pm i\sqrt{7}}{4}$	$1 \pm i$	6

Solve each equation by use of the Quadratic Formula. $x = \dfrac{-b \pm \sqrt{b^2-4ac}}{2a}$

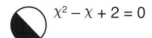 $x^2 - 3x + 4 = 0$

$x^2 - x + 1 = 0$

$x^2 - 2x + 2 = 0$

 $2x^2 + x + 1 = 0$ $x^2 - x + 2 = 0$ $3x^2 - x + 1 = 0$

Find the sum of the roots of each equation. Hint: $r_1 + r_2 = \dfrac{-b}{a}$

 $2x^2 - 3x + 2 = 0$ $3x^2 + 4x - 8 = 0$ $-5x^2 - 4x + 10 = 0$

Find the product of the roots of each equation. Hint: $r_1 \cdot r_2 = \dfrac{c}{a}$

 $4x^2 - x - 10 = 0$ $-3x^2 + 2x + 8 = 0$ $6x^2 - 5x - 10 = 0$

Write an equation in the form $ax^2 + bx + c = 0$ for the given roots.

 $r_1 = 4, \; r_2 = 3$ $r_1 = \dfrac{1}{4}, \; r_2 = -\dfrac{3}{4}$ $r_1 = 3 + i, \; r_2 = 3 - i$

Algebra II Topics by Design / Russell F. Jacobs   www.tessellations.com   © 2007 Jacobs Publishing Company

Name: _____

Solve for x in each exercise.

	A	B	C	D	E	F	
	-9	4	$3\frac{1}{2}$	$\frac{1}{4}$	-9	4	1
	-4	$-\frac{1}{3}$	1	3	-4	$-\frac{1}{3}$	2
	5	-1	-7	2	5	-1	3
	$-2\sqrt{2}$	-3	2	-7	$-2\sqrt{2}$	-3	4
	-9	4	$3\frac{1}{2}$	$\frac{1}{4}$	-9	4	5
	-4	$-\frac{1}{3}$	1	3	-4	$-\frac{1}{3}$	6

$3^{2x+1} = 3^{x+2}$

$2^{x-2} = 2^{2x-1}$

$5^{3x+2} = 5^{2x-5}$

$10^{-2x-3} = 10^{4x-1}$

$7^{2x+\sqrt{2}} = 7^{x-\sqrt{2}}$

 $2^{2x} \cdot 2^3 = 2^{x-1}$

Example: $27^x = 3^6$

$(3^3)^x = 3^6$

$3^{3x} = 3^6$

$3x = 6$

$x = 2$

$16^{x+1} = 2^5$

$25^{x-1} = 5^{x+3}$

$4^{x-1} = (\sqrt{2})^4$

$49^{2x-2} = 7^{3x}$

$81^{2x+3} = 3^{3x-3}$

$10^{x+3} = 1000^{-2}$

$\left(\frac{9}{16}\right)^2 = \left(\frac{3}{4}\right)^{x+1}$

$\left(\frac{1}{100}\right)^{x-2} = \left(\frac{1}{10}\right)^3$

Name: _____

	A	B	C	D	E	F	
1	32	−1	0.001	−2	5	−4	1
2	−4	5	16	10	−1	32	2
3	3	$\frac{1}{8}$	1000	7	3	$\frac{1}{8}$	3
4	$\frac{1}{4}$	25	8	9	$\frac{1}{4}$	25	4
5	5	−4	0.001	−2	32	−1	5
6	−1	32	16	10	−4	5	6

On the grid, A is the graph of $y = 2^x$. B is the graph of $x = 2^y$.
Answer the following questions.

 In B, what is the value of x if y equals 3?

 In B, what is the value of x if y equals −2?

 In A, what is the value of y if x equals 4?

 In A, what is the value of y if x equals −3?

Solve for x in each equation.

 $\log_2 x = 5$

 $\log_5 x = 2$

 $\log_{\frac{1}{2}} 16 = x$

 $\log_{\frac{1}{2}} 4 = x$

 $\log_x 49 = 2$

 $\log_x 0.1 = -1$

 $\log_x 27 = \frac{3}{2}$

 $\log_{\frac{3}{4}} \frac{4}{3} = x$

 $\log_{10} x = 3$

 $\log_4 64 = x$

 $\log_x 125 = 3$

$\log_{100} x = -\frac{3}{2}$

Name:_____

	A	B	C	D	E	F	
	$-\frac{5}{2}n+\frac{5}{2}$	11,325	$-5n+15$	36	$-\frac{5}{2}n+\frac{5}{2}$	36	1
	5,050	$3n-9$	$-5n+15$	11,325	$-5n+15$	$3n-9$	2
	20	-15	15	$-4n+11$	20	-15	3
	1,275	$3n-1$	-116	$5n-16$	1,275	$3n-1$	4
	$-\frac{5}{2}n+\frac{5}{2}$	36	$-5n+15$	36	$-\frac{5}{2}n+\frac{5}{2}$	11,325	5
	5,050	$3n-9$	5,050	11,325	5,050	$3n-9$	6

Determine the values of s_1 and d for each arithmetic sequence.

Then write an expression for the sequence based on $s_1 + (n-1)d$.

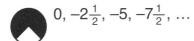 2, 5, 8, 11, …

7, 3, –1, –5, …

–6, –3, 0, 3, …

10, 5, 0, –5, …

$0, -2\frac{1}{2}, -5, -7\frac{1}{2}, …$

–11, –6, –1, 4, …

Use $S_n = \frac{n}{2}(s_1 + l)$ to find the sum of each arithmetic series.

$\sum\limits_{n=1}^{6} 2n - 1$

 $\sum\limits_{n=1}^{10} -n + 4$

$\sum\limits_{n=1}^{8} -3n - 1$

Use $S_n = \frac{n(n+1)}{2}$ to find the sum of each arithmetic series.

$1 + 2 + 3 + … + 100$

$1 + 2 + 3 + … + 50$

$1 + 2 + 3 + … + 150$

Use $l = s_1 + (n-1)d$ to find the number of terms in each arithmetic series.

 $-2 + 1 + 4 + 7 + … + 55$

 $12 + 8 + 4 + 0 + (-4) + … + (-44)$

Name:_____

	A	B	C	D	E	F	
1	520	126	−252	−5	520	126	
2	$5\frac{1}{3}$	2	−3	−183	0	2	
3	$\frac{1}{4}$	126	680	$-\frac{1}{2}$	$\frac{1}{4}$	27	
4	$28\frac{4}{5}$	−1	$-\frac{1}{2}$	680	$28\frac{4}{5}$	−1	
5	520	126	−252	−5	520	126	
6	0	2	−3	−183	$5\frac{1}{3}$	2	

Find the common ratio of each geometric sequence.

○ −4, 12, −36, 108, −324

◧ 8, −8, 8, −8, 8

▨ 100, −50, 25, $-12\frac{1}{2}$, $6\frac{1}{4}$

◔ −1, 5, −25, 125, −625

◕ −10, −20, −40, −80, −160

◣ 24, 6, $\frac{3}{2}$, $\frac{3}{8}$, $\frac{3}{32}$

Hint: Write the terms of the series first.

Example: $\sum\limits_{n=1}^{4} (-2)^n = -2 + 4 - 8 + 16$

Find the sum of each geometric series.

○ $\sum\limits_{n=1}^{5} (-3)^n$

◑ $\sum\limits_{n=1}^{6} 3(-2)^n$

◕ $\sum\limits_{n=1}^{10} 2(-1)^n$

◔ $\sum\limits_{n=1}^{6} -2(2)^n$

● $\sum\limits_{n=1}^{4} (-5)^n$

◨ $\sum\limits_{n=1}^{8} 4(-2)^n$

Use $S_n = \frac{s_1}{1-r}$ to find the limit of each geometric series.

 $8 - 4 + 2 - 1 + \frac{1}{2} + \ldots$

 $24 + 4 + \frac{2}{3} + \frac{1}{9} + \frac{1}{54} + \ldots$

 $36 - 12 + 4 - \frac{4}{3} + \frac{4}{9} + \ldots$

Name:_____

Find the determinant of each matrix.

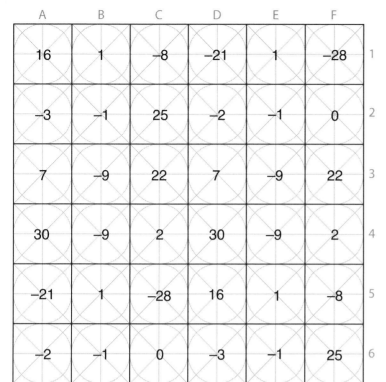

	A	B	C	D	E	F	
1	16	1	−8	−21	1	−28	1
2	−3	−1	25	−2	−1	0	2
3	7	−9	22	7	−9	22	3
4	30	−9	2	30	−9	2	4
5	−21	1	−28	16	1	−8	5
6	−2	−1	0	−3	−1	25	6

 $\begin{bmatrix} 5 & 3 \\ 1 & 2 \end{bmatrix}$

$\begin{bmatrix} -4 & -3 \\ 5 & 4 \end{bmatrix}$

$\begin{bmatrix} -1 & 2 \\ -3 & 4 \end{bmatrix}$

$\begin{bmatrix} \frac{1}{2} & 8 \\ \frac{1}{4} & 6 \end{bmatrix}$

$\begin{bmatrix} 0 & 1 \\ 3 & -2 \end{bmatrix}$

$\begin{bmatrix} -2 & -3 \\ -4 & -5 \end{bmatrix}$

$\begin{bmatrix} -4 & 4 \\ -3 & 5 \end{bmatrix}$

$\begin{bmatrix} \frac{1}{2} & 16 \\ \frac{1}{4} & -10 \end{bmatrix}$

$\begin{bmatrix} -8 & -4 \\ -6 & -5 \end{bmatrix}$

$\begin{bmatrix} 0 & 7 \\ 3 & 5 \end{bmatrix}$

$\begin{bmatrix} 24 & -\frac{1}{3} \\ 18 & \frac{2}{3} \end{bmatrix}$

$\begin{bmatrix} 0.3 & 50 \\ 0.8 & 40 \end{bmatrix}$

$\begin{bmatrix} 2 & 3 & 6 \\ 1 & 4 & 8 \\ 2 & 0 & 5 \end{bmatrix}$

$\begin{bmatrix} 1 & 4 & 6 \\ 3 & 2 & 5 \\ 1 & 4 & 3 \end{bmatrix}$

$\begin{bmatrix} 2 & -3 & 4 \\ 0 & 6 & -1 \\ 2 & -3 & 4 \end{bmatrix}$

Name:_____

Add. Then find the determinant
of each sum.

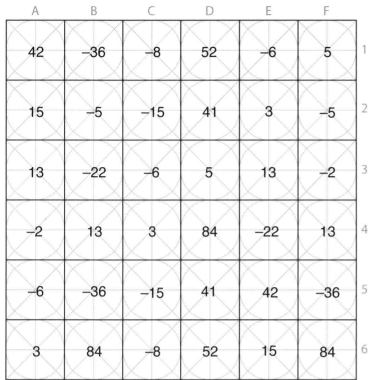

	A	B	C	D	E	F	
1	42	−36	−8	52	−6	5	
2	15	−5	−15	41	3	−5	
3	13	−22	−6	5	13	−2	
4	−2	13	3	84	−22	13	
5	−6	−36	−15	41	42	−36	
6	3	84	−8	52	15	84	

$\begin{bmatrix} -1 & -2 \\ 0 & 3 \end{bmatrix} + \begin{bmatrix} 2 & -1 \\ 2 & 4 \end{bmatrix}$

$\begin{bmatrix} 2 & -1 \\ 3 & 0 \end{bmatrix} + \begin{bmatrix} -5 & 4 \\ -1 & 3 \end{bmatrix}$

$\begin{bmatrix} -8 & 5 \\ 10 & -10 \end{bmatrix} + \begin{bmatrix} 6 & -3 \\ -7 & 8 \end{bmatrix}$

$\begin{bmatrix} 0 & -6 \\ 16 & -3 \end{bmatrix} + \begin{bmatrix} -5 & 4 \\ -8 & -2 \end{bmatrix}$

$\begin{bmatrix} -5 & 12 \\ 4 & 3 \end{bmatrix} + \begin{bmatrix} 0 & -7 \\ -7 & -3 \end{bmatrix}$

$\begin{bmatrix} -\frac{1}{2} & 5 \\ \frac{1}{4} & -1 \end{bmatrix} + \begin{bmatrix} 3\frac{1}{2} & -2 \\ \frac{3}{4} & 3 \end{bmatrix}$

$\begin{bmatrix} -8 & 2 \\ 15 & -1 \end{bmatrix} + \begin{bmatrix} 6 & -3 \\ -13 & 5 \end{bmatrix}$

$\begin{bmatrix} 3 & 2 & -1 \\ -2 & 0 & -2 \\ -\frac{3}{4} & -3 & 1 \end{bmatrix} + \begin{bmatrix} 1 & -3 & 2 \\ \frac{1}{2} & 4 & -1 \\ \frac{5}{4} & 1 & -2 \end{bmatrix}$

$\begin{bmatrix} 0 & 2 \\ -3 & -1 \end{bmatrix} + \begin{bmatrix} -6 & -5 \\ 7 & -4 \end{bmatrix}$

$\begin{bmatrix} -2 & 1 & -1 \\ 3 & 0 & -2 \\ -8 & 1 & 3 \end{bmatrix} + \begin{bmatrix} 5 & 2 & 4 \\ 1 & 4 & -1 \\ 4 & -1 & -6 \end{bmatrix}$

$\begin{bmatrix} 12 & -9 \\ 8 & 1 \end{bmatrix} + \begin{bmatrix} -15 & 7 \\ -10 & -4 \end{bmatrix}$

$\begin{bmatrix} 4 & -3 \\ -1 & 2 \end{bmatrix} + \begin{bmatrix} -8 & 6 \\ 3 & 2 \end{bmatrix}$

$\begin{bmatrix} 1 & 0 & 3 \\ 2 & -1 & 4 \\ -2 & 2 & 5 \end{bmatrix} + \begin{bmatrix} -5 & 2 & -1 \\ -2 & -2 & -3 \\ 4 & 1 & -3 \end{bmatrix}$

$\begin{bmatrix} 6 & -2 \\ 5 & 2 \end{bmatrix} + \begin{bmatrix} -4 & -2 \\ -9 & 2 \end{bmatrix}$

$\begin{bmatrix} -\frac{3}{4} & 0 \\ \frac{5}{2} & -3 \end{bmatrix} + \begin{bmatrix} \frac{5}{4} & 4 \\ -\frac{3}{2} & 1 \end{bmatrix}$

Name: _____

Multiply. Then find the determinant of the product.

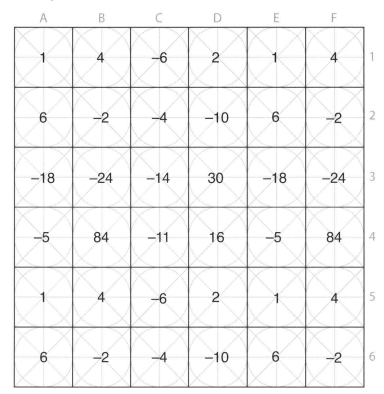

	A	B	C	D	E	F	
1	1	4	−6	2	1	4	
2	6	−2	−4	−10	6	−2	
3	−18	−24	−14	30	−18	−24	
4	−5	84	−11	16	−5	84	
5	1	4	−6	2	1	4	
6	6	−2	−4	−10	6	−2	

$\begin{bmatrix} 2 & -1 \\ 0 & 3 \end{bmatrix} \times \begin{bmatrix} -1 & 4 \\ -2 & 3 \end{bmatrix}$

$\begin{bmatrix} 1 & 3 \\ -4 & 2 \end{bmatrix} \times \begin{bmatrix} 2 & -1 \\ 0 & 3 \end{bmatrix}$

$\begin{bmatrix} -2 & 3 \\ 1 & -2 \end{bmatrix} \times \begin{bmatrix} 3 & -1 \\ 2 & -2 \end{bmatrix}$

$\begin{bmatrix} -2 & 3 \\ -1 & 1 \end{bmatrix} \times \begin{bmatrix} -1 & 5 \\ 3 & -4 \end{bmatrix}$

$\begin{bmatrix} -2 & 1 \\ 3 & -1 \end{bmatrix} \times \begin{bmatrix} 1 & 0 \\ 4 & 2 \end{bmatrix}$

$\begin{bmatrix} -2 & 0 \\ 0 & -1 \end{bmatrix} \times \begin{bmatrix} -3 & 2 \\ -2 & 1 \end{bmatrix}$

$\begin{bmatrix} 0 & 1 \\ -3 & -1 \end{bmatrix} \times \begin{bmatrix} 4 & -1 \\ -2 & 1 \end{bmatrix}$

$\begin{bmatrix} -1 & 2 \\ 3 & 0 \end{bmatrix} \times \begin{bmatrix} 2 & 3 \\ 0 & 2 \end{bmatrix}$

$\begin{bmatrix} 1 & -1 \\ -1 & -1 \end{bmatrix} \times \begin{bmatrix} 2 & -2 \\ -2 & -2 \end{bmatrix}$

$\begin{bmatrix} -3 & 2 \\ -2 & 1 \end{bmatrix} \times \begin{bmatrix} 0 & -2 \\ 2 & 3 \end{bmatrix}$

$\begin{bmatrix} -2 & -1 \\ -3 & 1 \end{bmatrix} \times \begin{bmatrix} 4 & -3 \\ 2 & -1 \end{bmatrix}$

$\begin{bmatrix} 3 & -4 \\ 0 & 2 \end{bmatrix} \times \begin{bmatrix} -1 & 0 \\ 2 & 3 \end{bmatrix}$

$\begin{bmatrix} -1 & 3 \\ 1 & -2 \end{bmatrix} \times \begin{bmatrix} 2 & 0 \\ -1 & 3 \end{bmatrix}$

$\begin{bmatrix} 1 & 0 \\ 0 & 1 \end{bmatrix} \times \begin{bmatrix} -1 & 1 \\ -2 & 1 \end{bmatrix}$

$\begin{bmatrix} 0 & 1 \\ 1 & 0 \end{bmatrix} \times \begin{bmatrix} -3 & -1 \\ 2 & -1 \end{bmatrix}$

$\begin{bmatrix} -3 & -2 \\ -2 & 1 \end{bmatrix} \times \begin{bmatrix} -2 & 0 \\ 3 & -1 \end{bmatrix}$

Name:_____

	A	B	C	D	E	F	
1	7	−3	$\frac{1}{8}$	6	7	−3	
2	$\frac{1}{3}$	−5	$-\frac{20}{7}$	4	$\frac{1}{3}$	−5	
3	$-\frac{18}{5}$	$\frac{18}{7}$	9	0	$-\frac{18}{5}$	$\frac{18}{7}$	
4	−7	3	5	$-\frac{7}{4}$	−7	3	
5	7	−3	$\frac{1}{8}$	6	7	−3	
6	$\frac{1}{3}$	−5	$-\frac{20}{7}$	4	$\frac{1}{3}$	−5	

Find the determinant of the coefficient matrix of each system of linear equations.

$$\begin{cases} x - 2y = 3 \\ x + y = 5 \end{cases}$$

$$\begin{cases} 2x - y = 4 \\ 3x + y = -2 \end{cases}$$

$$\begin{cases} 2x - y = 5 \\ 2x + y = -1 \end{cases}$$

$$\begin{cases} 3x - 1 = y \\ -x + 2 = 2y \end{cases}$$

$$\begin{cases} 3x + 2y = -1 \\ x - y = 4 \end{cases}$$

$$\begin{cases} \frac{1}{2}x + \frac{1}{4}y = 2 \\ 3x - 2y = -2 \end{cases}$$

$$\begin{cases} -x + 2y = 0 \\ 2x - y = -1 \end{cases}$$

$$\begin{cases} 4x - y - 1 = 0 \\ 2x + y + 4 = 0 \end{cases}$$

$$\begin{cases} y = -2x + 3 \\ y = 5x - 5 \end{cases}$$

$$\begin{cases} x + 2 = -2y \\ y - 1 = 4x \end{cases}$$

Use determinants to find the value of x in each system.

$$\begin{cases} -x + 4y = -2 \\ 2x - y = 5 \end{cases}$$

$$\begin{cases} -2x + y - 1 = 0 \\ x - 2y + 3 = 0 \end{cases}$$

$$\begin{cases} 3x - y + 2 = 0 \\ 2x + 2y - 5 = 0 \end{cases}$$

Use determinants to find the value of y in each system.

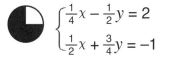

$$\begin{cases} \frac{1}{4}x - \frac{1}{2}y = 2 \\ \frac{1}{2}x + \frac{3}{4}y = -1 \end{cases}$$

$$\begin{cases} 3x - y = 3 \\ 2x + y = -4 \end{cases}$$

$$\begin{cases} -x + y - 1 = 0 \\ x - 2y + 1 = 0 \end{cases}$$

Name: _____

	A	B	C	D	E	F	
	$-x^3-3x^2$	x^2-2x+1	$2x^2+x-6$	-18	$x-2$	3	1
	1	$x-1$	$2x^2+2$	x^3-2x^2-x+2	$\frac{5}{4}$	-2	2
	$-3x+6$	$-2x+2$	14	$-3x+6$	$-2x+2$	14	3
	-4	$-2x+2$	-1	-4	$-2x+2$	-1	4
	-18	$x-2$	3	$-x^3-3x^2$	x^2-2x+1	$2x^2+x-6$	5
	x^3-2x^2-x+2	$\frac{5}{4}$	-2	1	$x-1$	$2x^2+2$	6

Remember $f(x)$ represents each y-value of function f.

Suppose $f(x) = x-4$.
Then $f(2) = -2$.
Thus, $(2, -2)$ is an ordered pair of the function.

Find each value.

 Let $f(x) = x^2-1$ and $g(x) = x+3$.
Find $f(-4) + g(-4)$.

 Let $f(x) = -x - x^2$ and $g(x) = x^3 + 5$.
Find $f(-1) - g(-1)$.

 Let $f(x) = x^2-4$ and $g(x) = x+1$, $x \neq -1$.
Find $f(3) \div g(3)$.

 Let $f(x) = -3$ and $g(x) = x^2 - x$.
Find $f(-2) \cdot g(-2)$.

Represent each value with an algebraic expression.

 Let $f(x) = x^3-1$
and $g(x) = -x^3+x$.
$f(x) + g(x) =$

 Let $f(x) = -2x^2+3$
and $g(x) = 4x^2-1$.
$f(x) + (gx) =$

 Let $f(x) = x^2-3$
and $g(x) = -x^2+ x + 1$.
$f(x) + g(x) =$

 Let $f(x) = x^3-2x^2$
and $g(x) = 2x^3+x^2$.
$f(x) - g(x) =$

 Let $f(x) = 5$
and $g(x) = 2x+3$.
$f(x) - g(x) =$

 Let $f(x) = x^2$
and $g(x) = 2x-1$.
$f(x) - g(x) =$

 Let $h(x) = x-2$
and $f(x) = -3$.
$h(x) \cdot f(x) =$

 $r(x) = 2x-3$
and $s(x) = x+2$.
$r(x) \cdot s(x) =$

 Let $w(x) = x^2-1$
and $t(x) = x-2$.
$w(x) \cdot t(x) =$

 Let $f(x) = |x-1|$
and $g(x) = \sqrt{x^2}$
Find $f(\frac{1}{4}) + g(\frac{1}{4})$.

 $[x]$ is the greatest integer less than or equal to x.
Find $[-2]$.

Find $[-\frac{3}{4}]$.

 Find $|-2| + [\frac{7}{4}]$.

Activity 41   Name: _____

	A	B	C	D	E	F	
	$y=-3x-1$	2	$y=2x+1$	$y=\frac{1}{2}(x+3)$	$y=2\sqrt{x+2}$	$(-3,3)$	1
	$y=\sqrt{x+5}$	$y=3x-18$	-1	$y=\frac{2}{5}(x+4)$	$(1,-1)$	$y=-x+\frac{1}{2}$	2
	$y=x^{\frac{1}{3}}$	$y=\log_b x$	$y=\frac{4}{x}$	$y=x^{\frac{1}{3}}$	$y=\log_b x$	$y=\frac{4}{x}$	3
	$(-1,0)$	$y=\log_b x$	$y=-2x+4$	$(-1,0)$	$y=\log_b x$	$y=-2x+4$	4
	$y=\frac{1}{2}(x+3)$	$y=2\sqrt{x+2}$	$(-3,3)$	$y=-3x-1$	2	$y=2x+1$	5
	$y=\frac{2}{5}(x+4)$	$(1,-1)$	$y=-x+\frac{1}{2}$	$y=\sqrt{x+5}$	$y=3x-18$	-1	6

Suppose function $f=\{(0,1), (1,-1), (-1,2)\}$. The **inverse** of f is $f^{-1}=\{(1,0), (-1,1), (2,-1)\}$. The ordered pairs of f^{-1} are obtained by reversing the elements in the ordered pairs of f. For example, $(1,0)$ is the **image** of $(0,1)$.

 $(-1,1)$ is an ordered pair in function h. What is its image in h^{-1}, the inverse of h?

 $(3,-3)$ is an ordered pair in function w. What is its image in w^{-1}, the inverse of w?

 $(0,-1)$ is an ordered pair in function g. What is its image in g^{-1}, the inverse of g?

Each equation below defines a function. Write an equation to define the inverse of the function.

Example: $y = 3x - 2$. Solution: $x = 3y - 2$. So, $y = \frac{1}{3}(x+2)$.

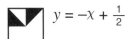 $y = 2x - 3$

 $y = x^2 - 5,\ x \geq 0$

 $y = -\frac{1}{2}x + 2$

$y = \frac{1}{3}(-x - 1)$

 $y = \frac{1}{3}x + 6$

$y = \frac{1}{4}x^2 - 2,\ x \geq 0$

 $y = \frac{5}{2}x - 4$

$y = \frac{1}{2}(x - 1)$

 $y = -x + \frac{1}{2}$

 $y = \frac{4}{x}$

 $y = x^3$

 $y = b^x,\ b > 0$

 In BASIC programming SGN(x) is the signum function.

$$\text{SGN}(x) = \begin{cases} 1 & \text{if } x > 0 \\ 0 & \text{if } x = 0 \\ -1 & \text{if } x < 0 \end{cases}$$

Find SGN($-3\frac{1}{2}$).

 In BASIC programming ABS(x) is the absolute value function. Find ABS(-2).

Activity 42

Name:_____

Find the missing term in each sequence.

	A	B	C	D	E	F	
1	$-x^5$	x^3	$11x^5$	i	$6+i$	x^3	1
2	$9\sqrt{2}$	$\dfrac{1}{x^5}$	$x^{\frac{5}{6}}$	i	$\sqrt{7}$	$\dfrac{1}{x^5}$	2
3	$9\sqrt{3}$	$-2x^2$	$2^{\frac{5}{2}}$	$-2x^2$	$9\sqrt{3}$	$-2x^2$	3
4	$5x-3$	$9\sqrt{3}$	$-4-4i$	$5x^{-5}$	$5x-3$	$9\sqrt{3}$	4
5	$6+i$	x^3	$11x^5$	i	$-x^5$	x^3	5
6	$\sqrt{7}$	$\dfrac{1}{x^5}$	$x^{\frac{5}{6}}$	i	$9\sqrt{2}$	$\dfrac{1}{x^5}$	6

$i, -1, -i, 1,$ ___?___

$\sqrt{3}, 3, 3\sqrt{3}, 9,$ ___?___

$2x^2, x^2, 0, -x^2,$ ___?___

$1 + i, 2, 2-2i, -4i,$ ___?___

$-x, x^2, -x^3, x^4,$ ___?___

$2x, 3x^2, 5x^3, 7x^4,$ ___?___

$\sqrt{3}, -\sqrt{4}, \sqrt{5}, -\sqrt{6},$ ___?___

$-2x^{-2}, 2x^{-1}, -2x^0, 2x^1,$ ___?___

$x, x\sqrt{x}, x^2, x^2\sqrt{x},$ ___?___

$\dfrac{1}{x}, x^{-2}, \dfrac{1}{x^3}, x^{-4},$ ___?___

$\sqrt{2}, \sqrt{6}, 3\sqrt{2}, 3\sqrt{6},$ ___?___

$x + 1, 2x, 3x - 1, 4x - 2,$ ___?___

$2^{\frac{1}{2}}, 2, 2^{\frac{3}{2}}, 4,$ ___?___

$2 + i, 3 - i, 4 + i, 5 - i,$ ___?___

$x^{-1}, -2x^{-2}, 3x^{-3}, -4x^{-4},$ ___?___

$x^{\frac{1}{2}}, x^{\frac{2}{3}}, x^{\frac{3}{4}}, x^{\frac{4}{5}},$ ___?___

Answer Key

Correct answers are given for each exercise. Activity One will be used as an example to show how each answer is represented. The first exercise in Activity One is in the upper right hand corner of the page. Its answer is $\frac{3}{2}$ and appears twice on the grid. One location on the grid is at the intersection of column D and row 4. So it can be represented as D4.

You can see D4 on the **Answer Key** in the approximate location of the exercise on the activity page. The answer $\frac{3}{2}$ can also be represented as C3. However, only one location will be shown for each correct answer.

Here is another example in Activity One. Look at the exercise in the lower left hand corner of the activity page. Its answer is $-\frac{3}{7}$ and can be represented as D5. It is shown on the **Answer Key** in the lower left hand corner in the same relative position as on the activity page.

ACTIVITY 1

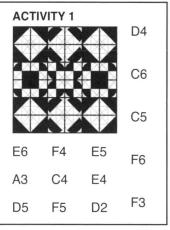

			D4
			C6
			C5
E6	F4	E5	F6
A3	C4	E4	
D5	F5	D2	F3

ACTIVITY 2

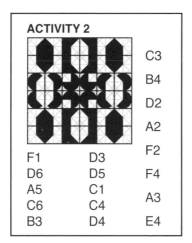

			C3
			B4
			D2
			A2
			F2
F1	D3		
D6	D5	F4	
A5	C1		A3
C6	C4		
B3	D4	E4	

ACTIVITY 3

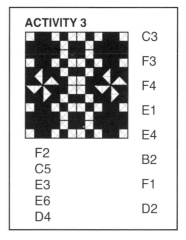

		C3
		F3
		F4
		E1
		E4
F2		B2
C5		
E3		F1
E6		D2
D4		

ACTIVITY 4

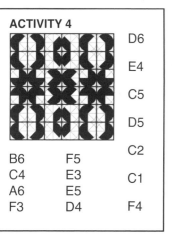

		D6
		E4
		C5
		D5
		C2
B6	F5	
C4	E3	C1
A6	E5	
F3	D4	F4

ACTIVITY 5

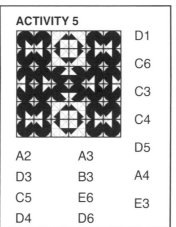

		D1
		C6
		C3
		C4
		D5
A2	A3	
D3	B3	A4
C5	E6	E3
D4	D6	

ACTIVITY 6

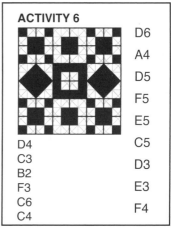

		D6
		A4
		D5
		F5
		E5
D4		C5
C3		D3
B2		
F3		E3
C6		F4
C4		

ACTIVITY 7

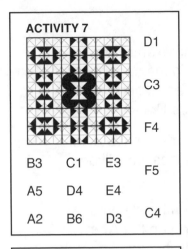

D1

C3

F4

B3	C1	E3	F5
A5	D4	E4	
A2	B6	D3	C4

ACTIVITY 8

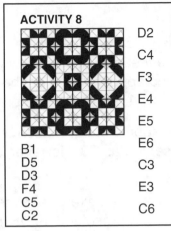

D2
C4
F3
E4
E5
E6

B1		
D5		C3
D3		
F4		E3
C5		
C2		C6

ACTIVITY 9

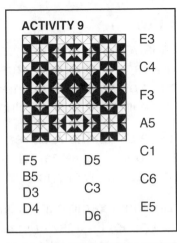

E3
C4
F3
A5
C1

F5	D5	
B5		C6
D3	C3	
D4	D6	E5

ACTIVITY 10

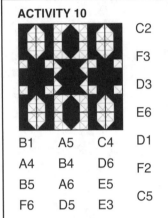

C2
F3
D3
E6

B1	A5	C4	D1
A4	B4	D6	F2
B5	A6	E5	
F6	D5	E3	C5

ACTIVITY 11

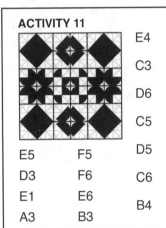

E4
C3
D6
C5

E5	F5	D5
D3	F6	C6
E1	E6	
A3	B3	B4

ACTIVITY 12

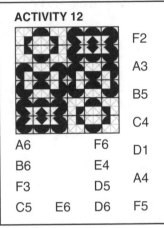

F2
A3
B5
C4

A6		F6	D1
B6		E4	
F3		D5	A4
C5	E6	D6	F5

ACTIVITY 13

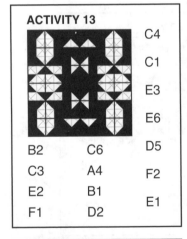

C4
C1
E3
E6
D5

B2	C6	
C3	A4	F2
E2	B1	
F1	D2	E1

ACTIVITY 14

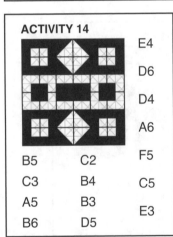

E4
D6
D4
A6
F5

B5	C2	
C3	B4	C5
A5	B3	
B6	D5	E3

ACTIVITY 15

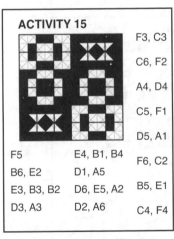

F3, C3
C6, F2
A4, D4
C5, F1
D5, A1

F5	E4, B1, B4	F6, C2
B6, E2	D1, A5	
E3, B3, B2	D6, E5, A2	B5, E1
D3, A3	D2, A6	C4, F4

ACTIVITY 16

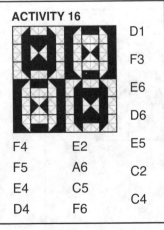

D1
F3
E6
D6
E5

F4	E2	
F5	A6	C2
E4	C5	
D4	F6	C4

ACTIVITY 17

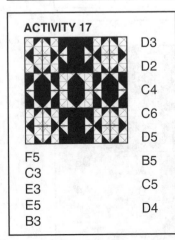

D3
D2
C4
C6
D5

F5		B5
C3		
E3		C5
E5		
B3		D4

ACTIVITY 18

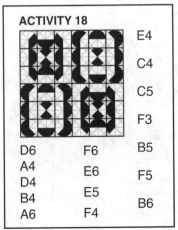

E4
C4
C5
F3
B5

D6	F6	
A4	E6	F5
D4		
B4	E5	
A6	F4	B6

ACTIVITY 19

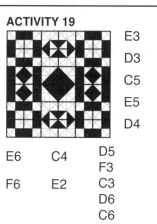

E3
D3
C5
E5
D4

E6   C4   D5
                F3
F6   E2   C3
                D6
                C6

ACTIVITY 20

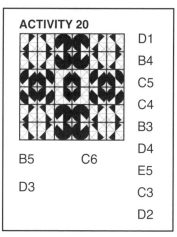

D1
B4
C5
C4
B3
D4

B5   C6   E5

D3   C3
                D2

ACTIVITY 21

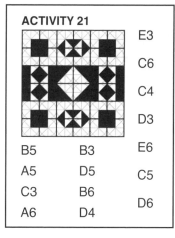

E3
C6
C4
D3
E6

B5   B3   C5
A5   D5
C3   B6   D6
A6   D4

ACTIVITY 22

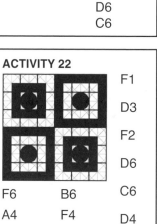

F1
D3
F2
D6
C6

F6   B6   D4
A4   F4
B4   F5   B5
D1   C1

ACTIVITY 23

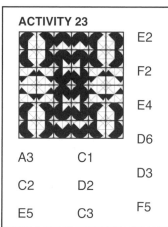

E2
F2
E4
D6

A3   C1
             D3
C2   D2
             F5
E5   C3

ACTIVITY 24

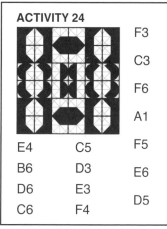

F3
C3
F6
A1

E4   C5   F5
B6   D3   E6
D6   E3
C6   F4   D5

ACTIVITY 25

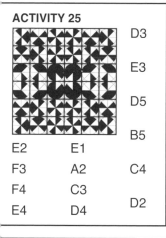

D3
E3
D5
B5

E2   E1
F3   A2   C4
F4   C3
E4   D4   D2

ACTIVITY 26

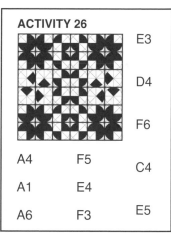

E3
D4
F6

A4   F5
             C4
A1   E4
A6   F3   E5

ACTIVITY 27

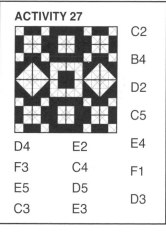

C2
B4
D2
C5
E4

D4   E2
F3   C4   F1
E5   D5
C3   E3   D3

ACTIVITY 28

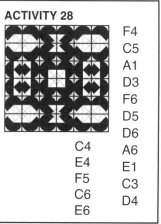

F4
C5
A1
D3
F6
D5
D6

    C4   A6
    E4   E1
    F5   C3
    C6   D4
    E6

ACTIVITY 29

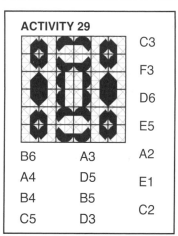

C3
F3
D6
E5
A2

B6   A3   E1
A4   D5
B4   B5   C2
C5   D3

ACTIVITY 30

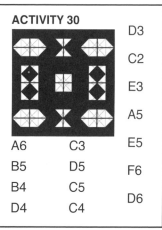

D3
C2
E3
A5
E5

A6   C3   F6
B5   D5
B4   C5   D6
D4   C4

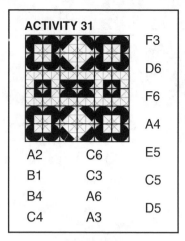

ACTIVITY 31

F3
D6
F6
A4
E5
C5
D5

A2	C6
B1	C3
B4	A6
C4	A3

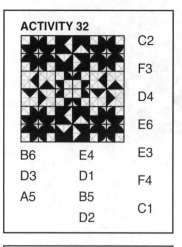

ACTIVITY 32

C2
F3
D4
E6
E3
F4
C1

B6	E4
D3	D1
A5	B5
	D2

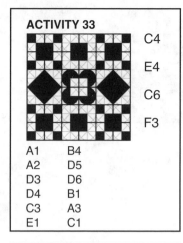

ACTIVITY 33

C4
E4
C6
F3

A1	B4
A2	D5
D3	D6
D4	B1
C3	A3
E1	C1

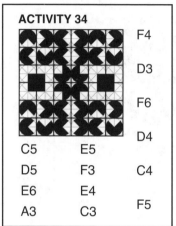

ACTIVITY 34

F4
D3
F6
D4
C4
F5

C5	E5
D5	F3
E6	E4
A3	C3

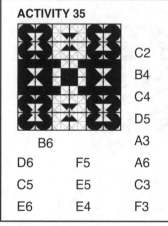

ACTIVITY 35

C2
B4
C4
D5
A3
A6
C3
F3

B6	
D6	F5
C5	E5
E6	E4

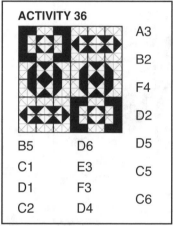

ACTIVITY 36

A3
B2
F4
D2
D5
C5
C6

B5	D6
C1	E3
D1	F3
C2	D4

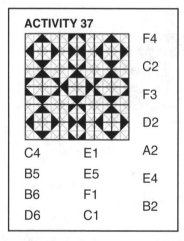

ACTIVITY 37

F4
C2
F3
D2
A2
E4
B2

C4	E1
B5	E5
B6	F1
D6	C1

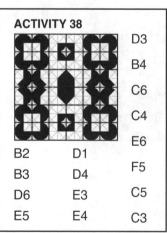

ACTIVITY 38

D3
B4
C6
C4
E6
F5
C5
C3

B2	D1
B3	D4
D6	E3
E5	E4

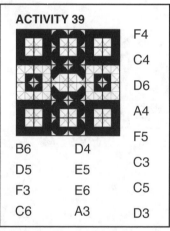

ACTIVITY 39

F4
C4
D6
A4
F5
C3
C5
D3

B6	D4
D5	E5
F3	E6
C6	A3

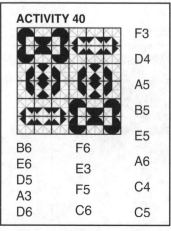

ACTIVITY 40

F3
D4
A5
B5
E5
A6
C4
C5

B6	F6
E6	E3
D5	F5
A3	
D6	C6

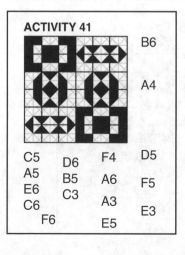

ACTIVITY 41

B6
A4
D5
F5
E3

C5	D6	F4
A5	B5	A6
E6	C3	A3
C6		E5
F6		

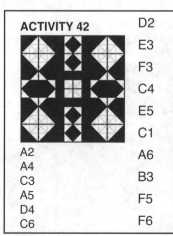

ACTIVITY 42

D2
E3
F3
C4
E5
C1
A6
B3
F5
F6

A2
A4
C3
A5
D4
C6